RECOMMENDED METHODS FOR PURIFICATION OF SOLVENTS AND TESTS FOR IMPURITIES

INTERNATIONAL UNION OF PURE AND APPLIED CHEMISTRY

Analytical Chemistry Division
Commission on Electroanalytical Chemistry

RECOMMENDED METHODS FOR PURIFICATION OF SOLVENTS AND TESTS FOR IMPURITIES

Edited by

J. F. COETZEE

University of Pittsburgh
Pennsylvania, USA

PERGAMON PRESS

OXFORD · NEW YORK · TORONTO · SYDNEY · PARIS · FRANKFURT

U.K.	Pergamon Press Ltd., Headington Hill Hall, Oxford OX3 0BW, England
U.S.A.	Pergamon Press Inc., Maxwell House, Fairview Park, Elmsford, New York 10523, U.S.A.
CANADA	Pergamon Press Canada Ltd., Suite 104, 150 Consumers Rd., Willowdale, Ontario M2J 1P9, Canada
AUSTRALIA	Pergamon Press (Aust.) Pty. Ltd., P.O. Box 544, Potts Point, N.S.W. 2011, Australia
FRANCE	Pergamon Press SARL, 24 rue des Ecoles, 75240 Paris, Cedex 05, France
FEDERAL REPUBLIC OF GERMANY	Pergamon Press GmbH, Hammerweg 6, D-6242 Kronberg-Taunus, Federal Republic of Germany

First edition 1982

Reprinted 1983

Library of Congress Cataloging in Publication Data

Main entry under title:
Recommended methods for purification of solvents and tests for impurities.
1. Solvents—Analysis. I. Coetzee, Johannes Francois.
QD544.R43 1982 541.3'482 82-15088

British Library Cataloguing in Publication Data

Recommended methods for purification of solvents and tests for impurities.
1. Electrochemistry
I. Coetzee, J.F. 2. International Union of Pure and Applied Chemistry. *Commission on Electro-analytical Chemistry*
541.3'72 QD533

ISBN 0-08-022370-2

In order to make this volume available as economically and as rapidly as possible the authors' typescripts have been reproduced in their original forms. This method unfortunately has its typographical limitations but it is hoped that they in no way distract the reader.

Printed in Great Britain by A. Wheaton & Co. Ltd., Exeter

CONTENTS

COMMISSION ON ELECTROANALYTICAL CHEMISTRY

Reports were originally published and subsequently revised during the period 1966 - 1981, when the following members served on the Commission on Electroanalytical Chemistry. (C: Chairman; S: Secretary; M: Member; AM: Associate Member; NR: National Representative;O: Observer).

J. Badoz-Lambling, O (France); R. G. Bates, AM and C (USA); B. Birch, NR (UK); E. Bishop, O, AM and M (UK); M. Branica, NR and AM (Yugoslavia); S. Bruckenstein, AM (USA); G. Charlot, NR, AM and M (France); J. F. Coetzee, AM, S and C (USA); A. K. Covington, AM and M (UK); P. Delahay, AM (USA); R. A. Durst, AM (USA); T. Fujinaga, NR, AM and M (Japan); Z. Galus, AM and M (Poland); L. Gierst, AM (Belgium); M. Gross, AM (France); K. Izutsu, AM and M (Japan); J. Jordan, NR, AM, M and S (USA); J. Juillard, AM and M (France); K. M. Kadish, AM (USA); R. Kalvoda, AM (Czechoslovakia); P. O. Kane, NR (UK); R. C. Kapoor, NR,AM and M (India); W. Kemula, NR, M and C (Poland); I. M. Kolthoff, M and C (USA); G. Kraft, NR (FRG); H. Laitinen, AM (USA), Y. Marcus, AM (Israel); J. Masek, AM (Czechoslovakia); L. Meites, AM and M (USA); T. Mussini, AM (Italy); R. Neeb, NR (FRG); H. W. Nürnberg, AM and M (FRG); B. Nygaard, AM (Sweden); P. Papoff, AM (Italy); D. D. Perrin, NR, AM and M (Australia); E. Pungor, AM and M (Hungary); W. C. Purdy, NR (Canada); R. A. Robinson, AM and S (USA); M. Senda, AM (Japan); D. Smith, AM (USA); W. F. Smyth, AM (UK); O. A. Songina, AM (USSR); J. Stradins, NR (USSR); N. Tanaka, NR, AM and M (Japan); J. K. Taylor, NR and AM (USA); K. Toth, NR (Hungary); B. Trémillon, AM and M (France); H. V. K. Udupa, NR (India); E. Vianello, M (Italy); A. A. Vlcek, AM (Czechoslovakia); P. Zuman, AM, M and S (USA).

PREFACE

The literature contains a proliferation of procedures of widely varying and frequently undocumented effectiveness recommended for the purification of important nonaqueous solvents. The result is that it is often difficult for workers to make a proper choice. The Commission on Electroanalytical Chemistry of the International Union of Pure and Applied Chemistry has been engaged for some time in a critical evaluation of such procedures by the Commission as a whole and aided by selected outside experts. The result has been the publication of recommended methods for ten solvents. These reports have received much attention worldwide from the many workers in different disciplines using such solvents.

The present compilation brings these ten reports, updated and revised where necessary, together in one volume. In addition, two introductory chapters dealing with general aspects of impurity effects are included in order to avoid needless duplication and also to strengthen the coherence of the subsequent reports which are written by a considerable number of authors. The two introductory chapters are followed by seven reports dealing with the important class of dipolar aprotic solvents for which impurity effects can be particularly severe. The first of these seven reports, that dealing with acetonitrile, also lays the groundwork for the next six chapters and has been totally rewritten in order to illustrate through specific examples (a) the general properties of the class of dipolar aprotic solvents, (b) typical impurity effects in such solvents, and (c) the typical chronology of improvements in purification procedures and tests for purity. The final three reports of the compilation deal with selected members of the class of amphiprotic solvents. The rationale for this particular selection of solvents is presented in Chapter 1. The Commission intends to publish further reports on additional important solvents.

It was thought inadvisable to force all reports into a rigid format. For example, as indicated above, the report on acetonitrile is constructed in such a way that it also serves as an introduction into the next six reports. On the other hand, the report on N-methylpropionamide contains an atypically detailed discussion of structural features of the liquid because the unusual properties of such mono-substituted amides are the result of these structural factors.

Finally, we anticipate that there will be a further continuous evolution in the effectiveness and also in the simplification of purification procedures and tests for purity of solvents. The Commission urges all workers to publish sufficient information on the purity of their solvents to make future interpretation of their data meaningful.

J. F. Coetzee, Editor*
University of Pittsburgh
December, 1981

---

* The Editor acknowledges financial support of his contributions to this compilation by the National Science Foundation.

CONTRIBUTORS

Meera Asthana
Chemistry Department
Faculty of Science
Banaras Hindu University
Varanasi 211 005
India

J. F. Coetzee
Department of Chemistry
University of Pittsburgh
Pittsburgh, Pennsylvania 15260
USA

T. Fujinaga
Department of Chemistry
Kyoto University
Sakyo-ku, Kyoto
Japan

T. B. Hoover
United States Environmental Protection Agency
Environmental Research Laboratory
Athens, Georgia 30601
USA

K. Izutsu
Department of Chemistry
Faculty of Science
Shinsu University
Matsumoto
Japan

Jean Juillard
Laboratoire d'Etude des Interactions Solutés-Solvants
Université de Clermont
BP 45, F-63170 Aubière
France

C. G. Karakatsanis
Mobay Chemical Corporation
Pittsburgh, Pennsylvania 15205
USA

L. A. Knecht
Department of Chemistry
Marietta College
Marietta, Ohio 45750
USA

M. W. Martin
Allied Chemical Company
Syracuse Research Laboratory
Solvay, New York 13209
USA

L. Mukherjee
Chemistry Department
Faculty of Science
Banaras Hindu University
Varanasi 221 005
India

T. B. Reddy
American Cyanamid Company
Chemical Research Division
Stamford, Connecticut 06904
USA

S. Sakura
Department of Chemistry
Faculty of Science
Kyoto University
Sakyo-ku, Kyoto
Japan

Solvent Purity: General Considerations

Prepared for publication by

J. F. Coetzee

The selection of appropriate solvents provides a powerful means to adjust both kinetic and thermodynamic features of the chemical reactivities of solutes over wide ranges. Only two examples will be given. (a) The rate constant of the S_N2 reaction

$$CH_3I + Cl^- \rightarrow CH_3Cl + I^-$$

is six powers of ten larger in the aprotic solvent dimethylformamide than in the protic solvent methanol (Ref. 1). (b) The range of accessible basicities can be extended by some eighteen powers of ten in lyate ion activity by replacing water with dimethyl sulfoxide as solvent (Ref. 2), while, at the other extreme, acidities can be increased by up to nine to twelve powers of ten in lyonium ion activity by replacing water with sulfolane as solvent (Ref. 3). The result of such simple control over solute reactivity is that many (if not most) reactions carried out in laboratories and in industrial processes are performed in solvents selected from a wide variety of different types (both organic and inorganic liquids, as well as fused salts) for such reasons as acquiring adequate solubility and stability of solutes, ensuring proper reaction paths in synthesis, and controlling reaction kinetics and thermodynamics in synthesis and analysis (Ref. 4-6).

The potential benefits of an appropriate choice of solvent may not actually be realized, however, if the solvent contains relatively reactive impurities. For example, all aprotic solvents have, by definition, only weak electrophilic properties. Consequently, many impurities (e.g., water) are sufficiently electrophilic to modify the properties of the medium significantly. On the other hand, certain dipolar aprotic solvents (e.g., acetonitrile, propylene carbonate and sulfolane) also have only weak nucleophilic properties (except when specific interactions occur, as between acetonitrile and copper(I) and silver(I) ions), so that many impurities (e.g., water in acetonitrile and sulfolane, and water and propylene glycols in propylene carbonate) are sufficiently nucleophilic to modify the properties of the medium significantly. It is therefore necessary to devote careful attention to the purity of the solvent, particularly if it is relatively inert, if reproducible and meaningful results are to be obtained. The recognition of this fact has led to such a proliferation of purification procedures and tests for impurities (real or merely potential) that it is difficult for the non-specialist to make a proper choice. The Commission on Electroanalytical Chemistry therefore undertook a number of years ago to solicit reports by specialists in which proposed purification procedures and tests for impurities would be critically evaluated. Ten such reports have been published; these are listed in Ref. 7-16.

The present compilation consists of relatively minor revisions of the reports listed in Ref. 13-16 and major revisions of the older reports listed in Ref. 7-11. The remaining older report, that listed in Ref. 12, remains essentially unchanged because there has been little further progress in the purification of N-methylacetamide.

The actual choice of solvents was dictated by their general usefulness as media for electrochemical measurements in particular (wide range of accessible potentials, ability to support radical ions and other reactive species, etc.). Consequently, the emphasis is on the class of dipolar aprotic solvents; but even so, the list is selective and illustrative, rather than exhaustive. For example, propylene carbonate is the only cyclic ether or ester included. Other members of the class, e.g., tetrahydofuran, dioxolane and γ-butyrolactone, and such solvents as N-methyl-2-pyrrolidone and dimethoxyethane are less important for electrochemical measurements but have found important applications in the new generation of aprotic battery systems; the purification of some of these solvents has been discussed by Butler et al. (Ref. 17).

TABLE 1. Properties and parameters of electrochemically useful solvents.

	W	NMAA[a]	NMPA	ED	AN	PC	SL[a]	DMF	DMSO	HMPA	PY
Freezing temperature, °C	0	+30	-30.9	+11.0	-45.7	-49.2	+28.45	-61	+18.55	+7.2	-41.7
Boiling temperature(1 atm.),°C	100	206	193	117.2	81.6	241.7	285	152.3	189.0	235	115.3
Vapor pressure, torr	23.76	1.5[b]	0.5[c]	10[d]	92			3.7	0.60	0.07[a]	10[e]
Dynamic viscosity, cp	0.894	3.89	5.2	1.54	0.344	2.53	10.3	0.796	2.00	3.24	0.884
Polarizability, $\mathring{A}^3$/molecule	1.47				4.45	8.56	11	7.91	7.97	18.8	
Dipole moment, D	1.76	4.4	3.6	1.90	4.1	4.9	4.7	3.9	4.1	5.4	2.37
Relative permittivity	78.30	179	175	12.9	36.0	64.9	43.3	37.0	46.7	29.8	12.3
Kosower Z[f]	(92)	77.9			71.3			68.5	70.2	62.8	64.0
Dimroth E_T[g]	63.1	52.0			46.0	46.6		43.8	45.0	40.9	40.2
Taft π^*[h]	1.09				0.76		0.95	0.875	1.000		
Gutmann DN[i]	(18)			55	14.1	15.1	14.8	26.6	29.8	38.8	33.1
Gutmann ACN[j]	54.8				19.3	18.3		16.0	19.3	10.6	14.2

Note: 1. Abbreviations for solvents are: W, water; NMAA, N-methylacetamide; NMPA, N-methylpropionamide; ED, ethylenediamine; AN, acetonitrile; PC, propylene carbonate; SL, sulfolane; DMF, N,N-dimethylformamide; DMSO, dimethyl sulfoxide; HMPA, hexamethylphosphoramide; PY, pyridine.
2. Values of temperature-dependent properties are for 25°C except where otherwise noted.
3. Principal sources of listed values of properties are in subsequent chapters in this compilation, in Ref. 18, and in Ref. 5, Vol. 2, chapters 9 and 14. Sources of values of polarity and solvation parameters are indicated below; also see Ref. 23 for critical comparison.

[a] Temperature is 30°C. [b] At 56°C. [c] At 60°C. [d] At 21.5°C. [e] At 13.2°C. [f] A proposed measure of the ionizing power of the solvent, based on the longest wavelength charge-transfer band of 1-ethyl-4-carbomethoxypyridinium iodide (Ref. 19). [g] Similar to Z, but based on a pyridinium N-phenoxide betaine (Ref. 20). [h] A measure of the polarity of the solvent, but based on the $\pi \rightarrow \pi^*$ transition of nitroaromatic indicators (Ref. 21). [i] A measure of the electron-donor strength of the solvent, based on its enthalpy of reaction with antimony pentachloride in 1,2-dichloroethane as solvent (Ref. 22). [j] A measure of the electron-acceptor strength of the solvent, based on the ^{31}P chemical shift of triethylphosphine oxide in the solvent (Ref. 22).

Relevant properties, as well as certain polarity and solvation parameters, of the solvents discussed here are listed in Table 1.

Certain general considerations apply to the purification of all solvents included here.

1. The nature and concentrations of impurities present in a given commercially produced solvent sometimes change, generally without notice. In some cases, the reason may be unannounced changes in manufacturing processes. Different manufacturers may use different processes. One example is the alarming inconsistencies found in the properties of acetonitrile obtained from different suppliers (Ref. 24). We recommend that, whenever possible, a given study be completed with solvent from one batch only, and that the source of the solvent, its batch number (if available), the purification procedure used and the results of tests for residual impurities always be reported.

2. It is seldom feasible, but it is fortunately rarely necessary, to lower the concentrations of all impurities to very low levels. The most reasonable strategy is usually to tailor the purification procedure to the intended use of the solvent.

3. The most generally applicable purification procedure is fractional distillation preceded, when really necessary, by deactivation of harmful impurities with appropriate reagents. However, the temptation must be resisted to use highly reactive reagents which may cause more harm than good, particularly if solvent decomposition occurs. There also has been a tendency (understandably) to make procedures needlessly complex. It is in fact probable that some of the procedures recommended in this compilation can be simplified for certain applications. On the other hand, for critical applications requiring ultrapure solvents, more use can perhaps be made of zone refining and preparative gas chromatography.

4. It is almost always necessary to devote special attention to the removal and prevention of subsequent uptake of the ubiquitous impurity, water. Fortunately, its reactivity is somewhat diminished by interaction with the solvent. At relatively low concentrations (of the order of a few millimolar) of water in aprotic solvents (S), the principal water species present are the monomer, W, and the two hydrogen bonded complexes WS and WS_2. The formation constants of such complexes are listed in Table 2, and the distribution of water among the states

TABLE 2. Stepwise formation constants of 1:1 and 1:2 hydrogen bonded complexes between water and various solvents at 30°C

Solvent	K_1	K_2	Solvent	K_1	K_2
Nitromethane	3.8	0.19	Dimethylformamide	43	0.92
Acetonitrile	4.1	0.17	Dimethyl sulfoxide	59	0.45
Pyridine	25	0.26	Hexamethylphosphoramide	446	2.5

Units: (mol fraction)$^{-1}$. Evaluated from PMR chemical shifts of 3×10^{-3} M water in binary mixtures of the indicated solvent and carbon tetrachloride (Ref. 25).

W, WS and WS_2 is shown in Figure 1 for the strong hydrogen bond acceptor hexamethylphosphoramide, and in Figure 2 for the weak acceptor acetonitrile. However, in spite of such stabilization of water by the solvent, the reactivity of water remains sufficiently high (particularly in solvents that are weak hydrogen bond acceptors) to modify the properties of the medium significantly. We therefore recommend the following for all measurements carried out in aprotic solvents and in which water may conceivably interfere: (a) The concentration of water must be lowered as far as practicable, (b) the concentration of residual water must be determined, preferably both before and after the measurement so as to also monitor water uptake, and (c) water must be deliberately added, its influence assessed and preferably extrapolated to zero water concentration. The same considerations apply to other reactive impurities.

5. Commonly used desiccants vary greatly in effectiveness, and their relative effectiveness is not the same in different solvents (26). Molecular Sieves (3A or 4A) is the most generally effective desiccant. As for distribution equilibria in general, shaking successively with several small batches of desiccant is more effective than using the same total amount of desiccant in one batch. Even more effective is flowing the solvent through a column packed with the desiccant. Flowing the solvent successively through columns packed with chromatographic alumina (Ref. 27) and Molecular Sieves is a generally effective means of lowering the content of water and other reactive impurities as well. The use of a Soxhlet extractor packed with the desiccant is particularly effective (but time consuming) for drying solvents and also solutions of sufficiently stable solutes (Ref. 28). Caution: We advise against drying solutions of perchlorates or other explosive materials in this way, as suggested in Ref. 28.

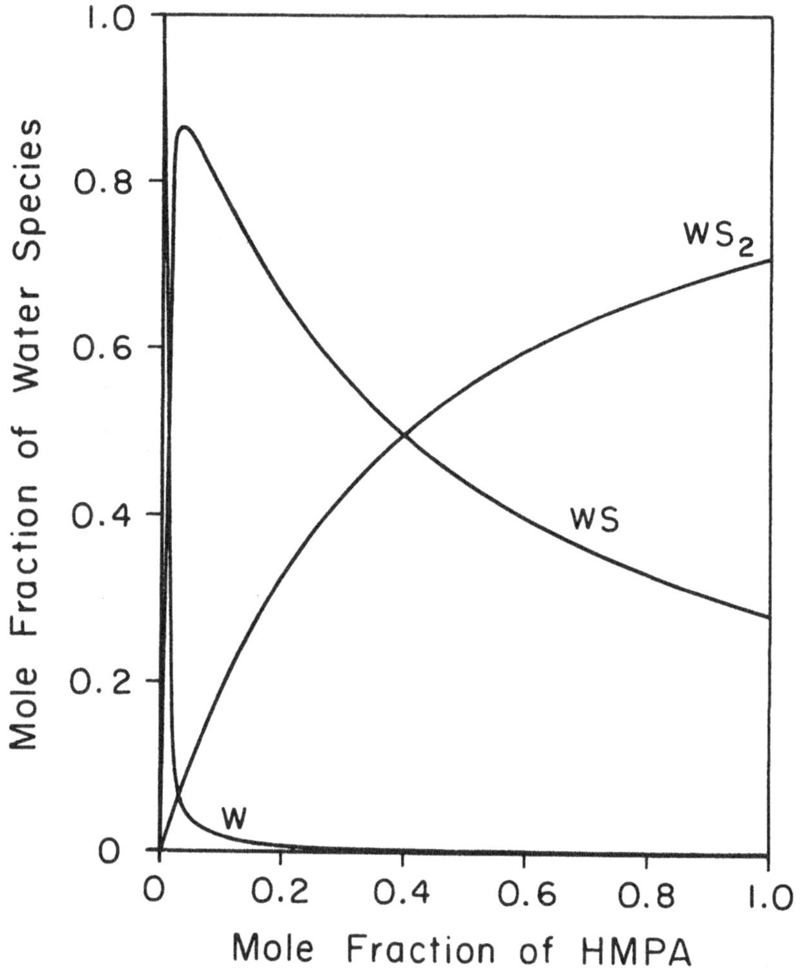

Figure 1. Distribution of 3×10^{-3} M water among the species W, WS, and WS_2 in binary mixtures of carbon tetrachloride and S = hexamethylphosphoramide (Ref. 25).

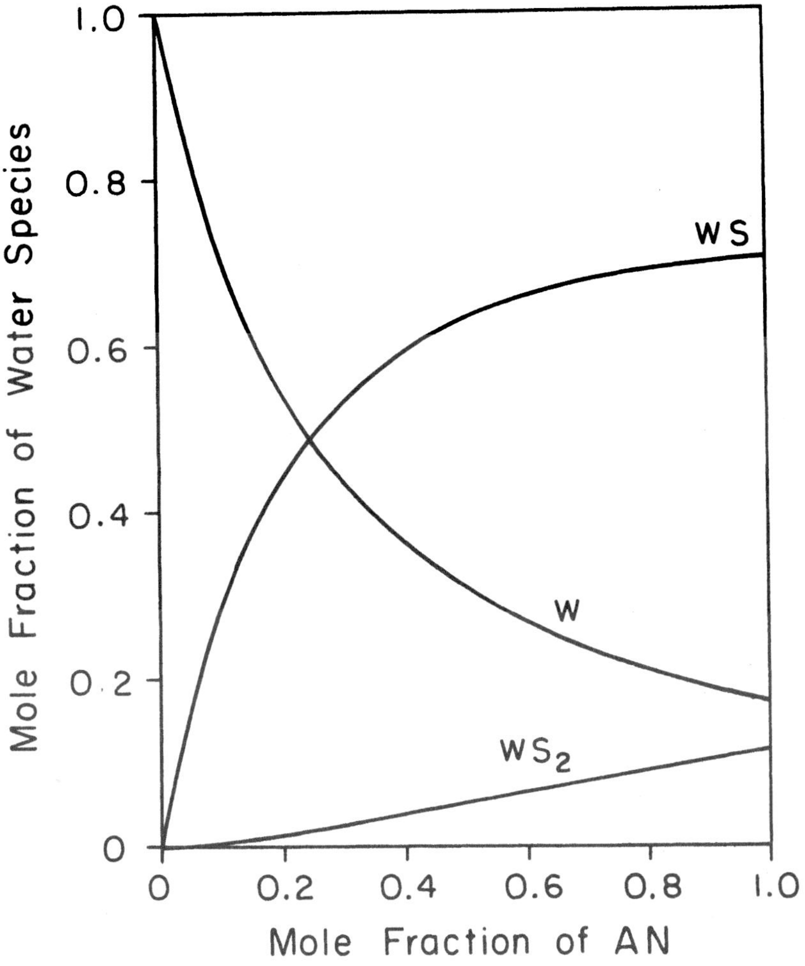

Figure 2. Distribution of 3×10^{-3} M water among the species W, WS, and WS_2 in binary mixtures of carbon tetrachloride and S = acetonitrile (Ref. 25).

Many solvents may be stored over Molecular Sieves, but solutions of electrolytes will undergo ion exchange with sodium or potassium ions present in the Sieves. It is generally best to store solvents and solutions in the dark under dry nitrogen or argon. Caution: If drying tubes are used to vent storage vessels to the atmosphere, a desiccant other than the commonly used but potentially explosive compound anhydrous magnesium perchlorate must be used; this compound has exploded violently in contact with acetone and may react similarly with other solvents.

6. While Karl Fischer titration is usually the method of choice for the determination of water, proton magnetic resonance (not very sensitive) and infrared spectrophotometry, as well as gas chromatography, can often provide the same information. The sensitivity of Karl Fischer titration (preferably performed coulometrically) is typically a few tenths of one millimolar water. There is a need for more sensitive methods. One such method is based on the reaction of water with phenylisocyanate, followed by determination of the resulting N,N'-diphenylurea by high performance liquid chromatography (Ref. 29).

7. The analytical methods of choice for the determination of other impurities in solvents are usually voltammetry, conductometry, ultraviolet and infrared spectrophotometry, and especially gas chromatography. Workers are just beginning to make use of gas chromatography-mass spectrometry; one example is the recent determination of previously unknown impurities in acetonitrile (Ref. 30). More use should be made of this powerful combination. If the solvent is intended for voltammetric experiments, differential pulse voltammetry at glassy carbon electrodes is particularly useful.

REFERENCES

1. A. J. Parker, Chem. Rev. 69, 1 (1969).
2. C. D. Ritchie and R. E. Uschold, J. Amer. Chem. Soc. 89, 1721, 2752 (1967).
3. J. F. Coetzee and R. J. Bertozzi, Anal. Chem. 45, 1064 (1973).
4. J. J. Lagowski, Ed., "The Chemistry of Nonaqueous Solvents," Academic Press, New York, 6 volumes, 1965-1978.
5. J. F. Coetzee and C. D. Ritchie, Eds., "Solute-Solvent Interactions," Marcel Dekker, New York, 2 volumes, 1968 and 1976.
6. A. K. Covington and T. Dickinson, Eds., "Physical Chemistry of Organic Solvent Systems," Plenum Press, London and New York, 1973.
7. L. M. Mukherjee and S. Bruckenstein, Pure Appl. Chem. 13, 421 (1966): ethylenediamine.
8. J. F. Coetzee, Pure Appl. Chem. 13, 429 (1966): acetonitrile.
9. T. B. Reddy, Pure Appl. Chem. 25, 459 (1971): dimethyl sulfoxide.
10. R. Lindauer and L. M. Mukherjee, Pure Appl. Chem. 27, 267 (1971): pyridine.
11. T. Fujinaga and K. Izutsu, Pure Appl. Chem. 27, 275 (1971): propylene carbonate.
12. L. A. Knecht, Pure Appl. Chem. 27, 283 (1971): N-methylacetamide.
13. T. B. Hoover, Pure Appl. Chem. 37, 581 (1974): N-methylpropionamide.
14. T. Fujinaga, K. Izutsu, and S. Sakura, Pure. Appl. Chem. 44, 117 (1975): hexamethylphosphoramide.
15. J. F. Coetzee, Pure Appl. Chem. 49, 213 (1977): sulfolane.
16. J. Juillard, Pure Appl. Chem. 49, 887 (1977): dimethyformamide.
17. J. N. Butler, R. J. Jasinski, D. R. Cogly, H. L. Jones, J. C. Synnott, and S. Carroll, "Purification and Analysis of Organic Nonaqueous Solvents," National Technical Information Service, Springfield, Va. 22151, U.S.A., Report AFCRL-70-0605, 1970.
18. J. A. Riddick and W. B. Bunger, "Organic Solvents," Wiley-Interscience, New York, 3rd edition, 1970.
19. E. M. Kosower, "An Introduction to Physical Organic Chemistry," Wiley, New York, 1968.
20. C. Reichardt, in A. D. Buckingham, E. Lippert, and S. Bratos, Eds., "Organic Liquids," Wiley, New York, 1978, pp. 269-291.
21. J. L. M. Abboud, M. J. Kamlet, and R. W. Taft, in R. W. Taft, Ed., "Progress in Physical Organic Chemistry," Interscience-Wiley, New York, Vol. 13, 1981, pp. 485-630.
22. V. Gutmann, "The Donor-Acceptor Approach to Molecular Interactions," Plenum Press, New York, 1978.
23. T. R. Griffiths and D. C. Pugh, Coord. Chem. Rev. 29, 129 (1979).
24. M. Walter and L. Ramaley, Anal. Chem. 45, 165 (1973).
25. J. F. Coetzee and A. Hussam, unpublished results (1981).
26. D. R. Burfield and R. H. Smithers, J. Org. Chem. 43, 3966 (1978).
27. M. L. Moskovitz, American Laboratory, 142 (1980).
28. P. Arthur, W. M. Haynes, and L. P. Varga, Anal. Chem. 38, 1630 (1966).
29. B. Björkqvist and H. Toivonen, J. Chromatography 178, 271 (1979).
30. L. Carlsen, H. Egsgaard, and J. R. Andersen, Anal. Chem. 51, 1593 (1979).

Relative Reactivities of Solutes in Different Solvents: Guidelines to Probable Impurity Effects

Prepared for publication by

J. F. Coetzee

The reactivities of many solutes depend greatly on the nature of the solvent. Consequently, a given impurity may significantly modify the reactivity of a given solute in a particular solvent, but not in other solvents. For example, even relatively low concentrations of water markedly increase the dissociation of weak Brønsted acids in acetonitrile, while in dimethylsulfoxide the influence of water is much less. The effect of impurities is particularly pronounced for ionic solutes, especially those ions in those solvents for which reactivities are high.

Thermodynamic aspects of relative reactivities of solutes in different solvents are represented by thermodynamic quantities for transfer of solutes from one solvent to another. Estimation of such quantities for individual ions requires an extrathermodynamic assumption. The relative merits of various such assumptions have been thoroughly debated (Ref. 1-3). The most generally accepted assumption at this time is that transfer energies of tetraphenylarsonium ion and tetraphenylborate ion are equal (Ref. 1-4). Representative values obtained in this manner are listed in Tables 1-3. Formamide is included in Table 1, because only limited data are available for N-methylacetamide and N-methylpropionamide (Chapters 11 and 12). The salient features of the data in Tables 1-3 will now be briefly indicated; a thorough discussion has been presented by Parker et al. (Ref. 1).

The main factor responsible for the trends in Table 1 is the varying influence of solutes on the uniquely developed structure of water (Ref. 6). This influence depends mainly on the charge and size of the solute. The smaller singly-charged ions, such as lithium and fluoride ions, (and particularly multicharged ions) act as net structure makers, while the larger singly-charged alkali metal and halide ions, as well as thiocyanate and perchlorate ions, are net structure breakers. However, large uncharged solutes and even ions if they contain hydrophobic moieties (e.g., Ph_4As^+, Ph_4B^- and R_4N^+ for R larger than C_2H_5) act as net structure makers, for reasons described in Ref. 6. Some consequences of these varying influences are the following. (a) All solvents listed in Table 1 are much less structured than water and therefore allow more net structure making by ions. One result is that these solvents are poorer solvators of most ions in the entropic sense than water is. However, another result is that the nonaqueous solvents tend to be better solvators in the enthalpic sense of many ions, particularly cations, than water is; this is true even for such weak donors as acetonitrile and propylene carbonate. The net result is the following order of overall solvating ability (in the free energy sense) towards most cations listed: HMPA > DMSO > DMF > SL,FA > W > AN,PC. (For abbreviations, see Chapter 1, Table 1). As expected, since the order of Lewis basicities depends on the nature of the reference acceptor, there are some reversals in the order of solvating abilities towards certain other acceptors; for example, towards several multicharged cations the order is HMPA > DMSO > DMF > W > AN > SL,PC (Ref. 1 and 4). (b) Small, "hard" anions (e.g., fluoride and chloride ions), even large anions if they have localized charge (e.g., carboxylates: see Table 3), and especially most multicharged anions require hydrogen bonding for stabilization and are therefore more weakly solvated in most nonaqueous solvents, particularly aprotic solvents, than in water. Important consequences are that in such solvents these anions tend to form stable complexes and ion pairs, that their ionic salts tend to have low solubilities, and that impurity effects on solutes containing such anions tend to be large. On the other hand, large, "soft" anions (e.g., iodide, perchlorate, and picrate ions) gain some stabilization enthalpically through dispersion interactions with highly polarizable solvents (see Chapter 1, Table 1), so that such ions tend to behave as relatively inert solutes and their salts tend to be relatively soluble in many nonaqueous solvents.

Such generalizations are useful in alerting investigators to the classes of solutes that are most susceptible to impurity effects.

TABLE 1. Standard free energies, enthalpies and entropies of transfer of ions from water to other solvents.

Ion	Formamide			Acetonitrile			Propylene carbonate		
	ΔG_t°	ΔH_t°	$-T\Delta S_t^\circ$	ΔG_t°	ΔH_t°	$-T\Delta S_t^\circ$	ΔG_t°	ΔH_t°	$-T\Delta S_t^\circ$
Li^+	-2.3	-1.3	-1.0	---	---	---	+5.7	+0.9	+4.8
Na^+	-1.9	-3.9	+2.0	+3.3	-3.1	+6.4	+3.6	-1.6	+5.2
K^+	-1.5	-4.0	+2.5	+1.9	-5.4	+7.3	+1.4	-5.0	+6.4
Rb^+	-1.3	-4.1	+2.8	+1.6	-5.5	+7.1	-0.7	-5.6	+4.9
Cs^+	-1.8	-4.1	+2.3	---	---	---	-2.9	-6.2	+3.3
Ag^+	-3.7	-5.4	+1.7	-5.2	-12.6	+7.4	+3.8	-3.0	+6.8
Ph_4As^+	-5.7	-0.1	-5.6	-7.8	-2.5	-5.3	-8.5	-3.6	-4.9
Cl^-	+3.3	+0.8	+2.5	+8.7	---	---	+9.0	+6.7	+2.3
Br^-	+2.7	-0.4	+3.1	+7.6	+2.0	+5.6	+7.1	+4.2	+2.9
I^-	+1.8	-1.8	+3.6	+4.5	-1.7	+6.2	+4.2	-0.2	+4.4
SCN^-	+3.4	-2.6	+6.0	+3.0	---	---	---	---	---
ClO_4^-	---	---	---	+1.1	---	---	---	---	---
Ph_4B^-	-5.7	-0.1	-5.6	-7.8	-2.5	-5.3	-8.5	-3.6	-4.9

Ion	Sulfolane			Dimethylformamide			Dimethyl sulfoxide		
	ΔG_t°	ΔH_t°	$-T\Delta S_t^\circ$	ΔG_t°	ΔH_t°	$-T\Delta S_t^\circ$	ΔG_t°	ΔH_t°	$-T\Delta S_t^\circ$
Li^+	---	---	---	-2.3	-7.7	+5.4	-3.5	-6.3	+2.8
Na^+	-0.7	-3.6	+2.9	-2.5	-7.9	+5.4	-3.3	-6.6	+3.3
K^+	-1.0	-6.0	+5.0	-2.3	-9.4	+7.1	-2.9	-8.3	+5.4
Rb^+	-2.1	-6.4	+4.3	-2.4	-9.0	+6.6	-2.6	-8.0	+5.4
Cs^+	-2.4	-5.9	+3.5	-2.2	-8.8	+6.6	-3.0	-7.7	+4.7
Ag^+	-0.9	-3.2	+2.3	-4.1	-9.2	+5.1	-8.0	-13.1	+5.1
Ph_4As^+	-8.5	-2.5	-6.0	-9.1	-4.7	-4.4	-8.8	-2.8	-6.0
Cl^-	+12.6	+6.2	+6.4	+11.0	+5.1	+5.9	+9.2	+4.5	+4.7
Br^-	+9.5	+2.9	+6.6	+7.2	+0.8	+6.4	+6.1	+0.8	+5.3
I^-	+4.9	+2.1	+7.0	+4.5	-3.3	+7.8	+2.2	-3.2	+5.4
SCN^-	---	---	---	+3.9	-2.4	+6.3	+2.0	---	---
ClO_4^-	---	---	---	+0.1	-5.4	+5.5	-0.3	-4.6	+4.3
Ph_4B^-	-8.5	-2.5	-6.0	-9.1	-4.7	-4.4	-8.8	-2.8	-6.0

From Parker et al. (Ref. 1). Based on the extrathermodynamic assumption that corresponding transfer energies of Ph_4As^+ and Ph_4B^- are equal. Values are on the molar scale in kcal mol^{-1} (1 cal = 4.18 J) and are for 298 K except in the case of sulfolane, for which values are for 303 K. The reliability of a given number depends on a variety of factors, including the validity of the extrathermodynamic assumption used (which is difficult to assess), and experimental problems which depend particularly on the reactivity of the ion and the solvent and the purity of the solvent. The reliability of reported transfer energies is being critically assessed by the Commission on Electroanalytical Chemistry and will be the subject of future reports.

TABLE 2. Recommended values of free energies of transfer of ions from water to hexamethylphosphoramide.

Ion	H^+	K^+	Ag^+	Ph_4As^+	Cl^-	Br^-	I^-	SCN^-	ClO_4^-	Ph_4B^-
ΔG_t°	-9.8	-3.6	-10.5	-9.3	+13.9	+11.0	+7.2	+4.8	-1.7	-9.3

From compilation by Marcus (Ref. 3). Details are the same as for Table 1, but recommended values were chosen from various sources according to criteria presented in Ref. 3.

TABLE 3. Free energies of transfer of ions from water to other solvents.

Ion	Formamide	Acetonitrile	Dimethylformamide	Dimethyl sulfoxide
Hydrogen	---	+11	-3.4	-4.5
Fluoride	+6.0	>17[a]	>14	---
Acetate	+4.4	+13	+16	+14
Benzoate	---	+10	+11	---
Picrate	---	-0.6	-2.2	+1.0

[a]Ref. 5

From compilation by Popovych (Ref. 2) except where otherwise noted. Details are the same as for Table 1, but uncertainties are likely to be greater in most cases; see text.

REFERENCES

1. B. G. Cox, G. R. Hedwig, A. J. Parker, and D. W. Watts, Australian J. Chem., 27 477 (1974).

2. O. Popovych, in I. M. Kolthoff and P. J. Elving, Eds., "Treatise on Analytical Chemistry," Wiley, New York, Part I, Vol. 1, 2nd ed., 1978, pp 711-771.

3. Y. Marcus, Rev. Anal. Chem. 5, 53 (1980).

4. J. F. Coetzee, Proc. Electrochem. Soc. 80-7, 70 (1980).

5. J. F. Coetzee and M. W. Martin, Anal. Chem. 52, 2412 (1980).

6. H. S. Frank and W. Y. Wen, Discuss. Faraday Soc. 24, 133 (1957).

Acetonitrile

Prepared for publication by

J. F. Coetzee and M. W. Martin

INTRODUCTION

Nitriles can be prepared by numerous methods (Ref. 1 and 2), but commercial acetonitrile is now usually (but not always: see Ref. 3) a byproduct of the large-scale production of acrylonitrile from propylene and ammonia. Acetonitrile is one of the most important members of the class of dipolar aprotic solvents. It is used extensively as a reaction medium in synthesis and for mechanistic studies, as well as in electrochemistry, spectrophotometry and liquid chromatography. Numerous examples of its uses are given in Ref. 4 and 5, and in Ref. 4-6 of Chapter 1. It owes its manifold applications in such diverse fields to a fortuitous combination of properties, particularly a high polarity (and permittivity) accompanied by only modest "chemical" solvating power towards many polar and ionic solutes. In common with all other dipolar aprotic solvents, acetonitrile typically behaves as a very weak electrophile. In common with such dipolar aprotic solvents as propylene carbonate and sulfolane, but unlike hexamethylphosphoramide, dimethyl sulfoxide and dimethylformamide, it is also a relatively weak nucleophile towards most (but not all) acceptors. Consequently, many solutes have high and strongly differentiated reactivities (both kinetically and thermodynamically) in acetonitrile. However, many of the potential benefits of acetonitrile as a reaction medium are not fully realized unless the solvent is carefully purified, stored and handled.

Other properties contributing to the usefulness of acetonitrile are its convenient liquid range, low viscosity, only moderate toxicity when pure, ready availability and inexpensiveness.

Values for selected properties of acetonitrile are listed in Table 1.

TABLE 1. Selected properties of acetonitrile

Property	Value	Ref.
Freezing temperature, °C	-45.7	6
Boiling temperature (1 atm.), °C	81.6	6
Flash point (closed vessel), °C	7.2	7
Vapor pressure, torr	92	6
Dynamic viscosity, cp	0.344	8
Density, g cm^{-3}	0.7766	7
Dipole moment, D	4.1	6
Relative permittivity	36.0	9
Refractive index, n_D	1.34163	7
Polarizability, $Å^3$/molecule	4.45	10
Surface tension (20°C), dyne cm^{-1}	19.10	7
Enthalpy of vaporization, kcal mol^{-1}		
-at 25°C	7.7	6
-at normal b.t.	7.1	6
Entropy of vaporization, cal K^{-1} mol^{-1}		
-at normal b.t.	20.06	6

Values of temperature-dependent properties are for 25°C, except where otherwise noted. For values of polarity and solvation parameters, see Chapter 1, Table 1.

STRUCTURE OF LIQUID ACETONITRILE

The structure of liquid nitriles has been the subject of controversy because such different probes as thermodynamic, dielectric and optical measurements have not always lead to the same conclusions. For example, Saum (Ref. 11) compared enthalpies of vaporization and other properties of nitriles with corresponding properties of alkanes and esters and, assuming specific antiparallel "dipole pair bonding", estimated for liquid nitriles an enthalpy of association of -8 kcal mol^{-1}. On the other hand, Dannhauser and Flueckinger (Ref. 12) concluded from dielectric data that the free energy of association might be as small as 1 kcal mol^{-1}. However, the more recent results of infrared investigations by Michel and X-ray analysis by Kratochwill, as well as other evidence reviewed by Michel and Lippert (Ref. 13), showed not only the expected antiparallel alignment of dipoles, but also a less stable axial orientation, which indicates that the positive charge of the dipole is located at the end of the molecule rather than being restricted to the nitrile carbon atom. (The approximate molecular orbital calculations of Pople and Gordon (Ref. 14) also show considerable positive charge on the hydrogen atoms.) Consequently, the centers of the dipole and of the molecule are similarly placed. There is almost totally free rotation around the axis of the molecule, and the potential around the axis is highly symmetrical. The environment of an individual molecule is lattice-like; this must influence reactivities of solutes, although apparently not as profoundly as the structure of water influences reactivities of solutes in aqueous solution (see Chapter 2).

The π-electrons of the nitrile triple bond can be polarized by electric fields in both axial and perpendicular directions (Ref. 13). Lewis acids can form axial σ-complexes with the lone pair of electrons, and they can also form perpendicular π-complexes. The direction of the infrared frequency shift produced by such complexation is towards higher values and is therefore atypical (Ref. 13, 15 and 16). The reason is that the lone pair electrons have partial antibonding character for the triple bond, so that the force constant is increased by Lewis interactions, thereby compensating for the normal decrease in frequency produced by intermolecular forces (Ref. 13). Modified CNDO calculations by Kecki and Wojtczak (Ref. 15) showed that the triple bond frequencies are strongly influenced by the lone pair electrons, while the intensities are influenced independently by polarization of the π-electrons. Lewis bases, on the other hand, cause the expected decrease in the frequencies of the C-H bonds (Ref. 13 and 16).

KEY SOLVENT PROPERTIES OF ACETONITRILE

It is beyond the scope of this compilation to discuss in any detail the properties of solutions in the ten solvents included. Nevertheless, in this chapter the most general properties of electrolytes in acetonitrile will be briefly mentioned in order to illustrate further the generalizations presented in Chapters 1 and 2 with special emphasis on the need for solvent purity, and also to serve as an entree into the extensive literature of the class of dipolar aprotic solvents.

1. Acetonitrile is an excellent solvent for many covalent compounds and for certain types of ionic compounds. For reasons indicated in Chapter 2, the most soluble salts are generally those containing symmetrical, large, polarizable and singly-charged anions, such as iodides, thiocyanates, perchlorates, tetrafluoroborates, hexafluorophosphates, tetraphenylborates and picrates. Solutions of ionophores containing such anions typically have good electrical conductivities.

2. The reactivities of solutes are typically strongly differentiated in acetonitrile. For example, acids and bases show a rich diversity in properties which is masked in more reactive solvents. The effective (operational) autoprotolysis constant of carefully purified and handled acetonitrile is less than 10^{-32} mol^2 L^{2-} (Ref. 17). In practice, the effective value will be larger due to the presence of impurities; even trace amounts of electrophilic impurities in particular, such as water and carbon dioxide, cause large increases (Ref. 18). Nevertheless, with due care a wide range of proton activities is accessible in acetonitrile.

3. A key property of those Brønsted acids forming anions that are poorly solvated by acetonitrile (e.g., chloride ion and most carboxylate and phenolate ions) is that dissociation occurs by either or both of the following overall pathways

$$HA + HA + CH_3CN \rightleftarrows CH_3CNH^+ + AH...A^-$$

$$HA + HX + CH_3CN \rightleftarrows CH_3CNH^+ + XH...A^-$$

whereby the high-energy anion A^- derives stability from hydrogen bonding by the parent acid HA ("homoconjugation") or by other protic compounds (e.g., water) present ("heteroconjugation"). Analogous but less stable complexes $BH^+...B$ are formed by cationic acids BH^+. Homoconjugation occurs in all dipolar aprotic solvents, but it is most pronounced in the weak nucleophiles acetonitrile, propylene carbonate and sulfolane, as expected. From the analytical point of view, homoconjugation and heteroconjugation are important because they affect the solubilities of many salts, the proton activities of buffer solutions, and the shapes of conductometric and potentiometric titration curves; this field has been extensively studied, particularly by Kolthoff and his associates, and has been reviewed in Ref. 4 and 5. Both the widespread occurrence of homoconjugation in acetonitrile and the differentiating nature of this solvent are illustrated in Table 2.

TABLE 2. Formation constants of homoconjugated complexes (K_f) and acid dissociation constants (K_a) of Brønsted acids in acetonitrile.

Acid	$\log (K_f)_{AN}$	$-\log (K_a)_{AN}$	$-\log (K_a)_W$	ΔpK_a
Perchloric	a	<2	---	---
Hydrobromic	2.4	5.5	---	---
Hydrochloric	2.2	8.9	---	---
o-Hydroxybenzoic	3.3	16.7	3.0	13.7
Benzoic	3.6	20.7	4.2	16.5
2,4,6-Trinitrophenol	a	11.0	0.7	10.3
Phenol	5.8	26.6	10.2	16.6
Ammonium ion	1.0	16.5	9.2	7.3

Data of Kolthoff *et al*. and Coetzee *et al*., selected from the compilation in Ref. 4. Obtained by a combination of solubility, conductometric, potentiometric (glass electrode) and spectrophotometric methods.

[a] Insignificant homoconjugation, owing to delocalized charge.

4. A specific practical result of the high reactivities of certain anions in acetonitrile and other dipolar aprotic solvents is that reference electrodes of the type Ag,AgCl/(Cl^- soln) or Hg, Hg_2Cl_2/(Cl^- soln) cannot be used in such solvents because the following reactions occur:

$$AgCl\downarrow + Cl^- \rightleftarrows AgCl_2^-$$

$$Hg_2Cl_2\downarrow + Cl^- \rightleftarrows Hg\downarrow + HgCl_3^-$$

Such reactions are promoted in dipolar aprotic solvents by the greater ability of the solvent to stabilize the negative charge in the highly polarizable complexes $AgCl_2^-$ and $HgCl_3^-$ than in the "hard" ion Cl^-.

5. Drawbacks of acetonitrile include its relatively high volatility (see Table 1) and its less-than-ideal long-term stability towards certain reagents, such as sulfuric acid and other relatively strong acids. It is necessary to guard against its hydrolysis to the much stronger electrophile and nucleophile acetamide, a reaction which is catalyzed by such Lewis acids as mercury(II) ion (Ref. 19). Finally, while pure acetonitrile is apparently only moderately toxic (Ref. 7), the impure solvent typically contains acrylonitrile which is highly toxic.

PURIFICATION OF ACETONITRILE

Different uses of the relatively inert solvent acetonitrile require different criteria for acceptable purity. The result is that the purification of this solvent has challenged many workers, particularly during the past thirty years. The chronology and key features of this development are presented in Table 3. During this period of time, there have been marked changes in the nature and concentrations of impurities; there have also been improvements in the detection of impurities and in their removal. The purification procedures recommended fifteen years ago by the Commission on Electroanalytical Chemistry (Ref. 29) are now being modified as described below; these procedures appear to be suitable for the purification of acetonitrile currently supplied by a variety of sources.

TABLE 3. Objectives, strategies and effectiveness of different purification procedures for acetonitrile.

Authors	Ref.	Year	Source[a]	Principal reagents added	Purpose[b]	Results[c]
Kolthoff, Coetzee	20	1956	MPG	Na_2CO_3; P_2O_5	Acetic acid; water	$C_W < 30$.
Coetzee, Cunningham, McGuire, Padmanabhan	21	1962	EPG	KOH; P_2O_5; CaH_2	Acrylonitrile; ammonia, other bases; water, other acids	$C_W < 20$, $\kappa < 10^{-7}$; leaves aromatics.
Forcier, Olver	22	1965	EPG, FPG, UCPG	NaH; P_2O_5; CaH_2	As for Coetzee *et al.*	As for Coetzee *et al.*
O'Donnell, Ayres, Mann	23	1965	MPG	Benzoyl chloride; $KMnO_4$ + Na_2CO_3; H_2SO_4	Acrylonitrile; aromatics; ammonia	$C_W < 100$, UV = 189.
McClure, Maricle	24	1967	BRG	$KMnO_4$ + Na_2CO_3; H_2SO_4; P_2O_5	Aromatics; ammonia; water	$C_W < 40$.
Moe	25	1967	Unspecified	Ethanol	Acrylonitrile, as azeotrope	Leaves two impurities (GLC).
Sherman, Olson	26	1968	MNG, MPQ	N_2O_4; CaH_2	Bases; water, other acids	$C_W < 20$.
Walter, Ramaley	3	1973	EPG, MPG, FRG	$AlCl_3$; $KMnO_4$ + Li_2CO_3; $KHSO_4$; CaH_2	Acrylonitrile; aromatics; bases; water, other acids	$\kappa = 5 \times 10^{-8}$, UV < 200, $C_W < 10$; acrylonitrile: C = 20. (for MPG and FRG only).
Carlsen, Egsgaard, Andersen	27	1979	Ugelor; other[d]	Na octanolate; H_2SO_4	Acrylonitrile; allyl alcohol	Acrylonitrile, allyl alcohol, benzene: C < 0.4, 0.6, 0.04; $C_W < 100$.
Kiesele	28	1980	Unspecified	Methanol; NaH; Al_2O_3; CaH_2	Acetone, benzene, as azeotropes; acrylonitrile; allyl alcohol; other acids	$C_W = 1.0$; other impurities: similar to Carlsen *et al.*

[a] EPG, FPG, MPG, UCPG: Eastman, Fisher, Matheson, Coleman and Bell, and Union Carbide practical grade; BRG: Baker reagent grade; FRG: Fisher reagent grade; MNG: Mallinckrodt "nanograde"; MPQ: Matheson, Coleman and Bell pesticide quality.

[b] Removal of indicated impurity; usually involves distillation.

[c] C: concentration of residual impurity, ppm; C_W = concentration of residual water, ppm (18 ppm = 1.0×10^{-3} M water); UV: ultraviolet cutoff (absorbance = 1 in 1-cm cell), nm; κ: electrical conductivity, $ohm^{-1}\ cm^{-1}$.

[d] Merck zur Synthese, Fluka purum (private communication from J. R. Andersen).

Procedure I (Ref. 28)

Step 1. Add 300 mL of methanol to 3 L of acetonitrile and fractionate at high reflux ratio until the boiling temperature rises from 64 to 80°C and the distillate becomes optically transparent down to 240 nm (disappearance of the "benzene fingers"). Discard the distillate (typically 500-550 mL).

Step 2. Add 1 g of sodium hydride suspension in paraffin oil to the remaining solvent, reflux for 10 minutes and then distill rapidly until *ca.* 100 mL of residue (yellow or brownish) remains.

Step 3. Immediately pass the distillate through a 3-cm diameter column packed with 300 g of acidic alumina (e.g., Woelm W 200 Super 1, Acid). Discard the first 150 mL of percolate, or add it to another batch of solvent intended for purification. The main fraction of the percolate is suitable for many uses, but has limited long-term stability, possibly due to the presence of colloidal alumina, which is removed in step 4.

Step 4. Add 5 g of calcium hydride and distill the first 50 mL at a high reflux ratio; discard this forecut. Finally, distill the main fraction, rapidly if desired.

Procedure II (Ref. 27)

Step 1. Add 300 mL of 0.3 M solution of sodium n-octanolate in n-octanol to 3 L of acetonitrile, reflux for 2 hours, and then distill quickly at a rate of *ca.* 500 mL per hour until *ca.* 500 mL of residue (yellow-brown) remains.

Step 2. Add 25-30 mL of concentrated sulfuric acid to the distillate and fractionate at a high reflux ratio. Best fractionation is obtained by maintaining vigorous boiling and collecting *ca.* 50 mL per hour of distillate by means of an adjustable reflux ratio head. Discard the first 15% forecut, which is rich in benzene.

Step 3. Immediately before use, pass the distillate through a column packed with 250 g of alumina (e.g., Woelm Super 1, Neutral) per L of solvent.

Procedure III (Ref. 30)

Step 1. Pass 3 L of acetonitrile successively through two 2-cm columns packed with 200 g of 3A or 4A Molecular Sieves and 200 g of neutral chromatographic alumina.

Step 2. Add 5 g of calcium hydride and fractionate at a high reflux ratio until the boiling temperature stabilizes at 81.6 °C (1 atm.). Discard the forecut and collect 90% of the remaining solvent, by faster distillation if desired.

Procedure III is not as effective overall as Procedures I and II, but it does yield a product from technical grade solvents which is low in water and other electrophilic impurities and which is suitable for many uses (not for voltammetry at very positive potentials or spectrophotometric work much below 230 nm, depending on the source). Measures of effectiveness are compared for Procedures I and II in Table 4. Neither procedure has much effect on

TABLE 4. Comparison of effectiveness claimed for two purification procedures.

Impurity	Procedure II[a]		Procedure I[b]	
	Before	After	Before	After
Water	---	$<1 \times 10^{2}$[c]	1.7×10^{3}	1.0[d]
Propionitrile	3.5×10^{2}	3.5×10^{2}	3.2×10^{2}	2.5×10^{2}
Acrylonitrile	1.3×10^{2}	$<4 \times 10^{-1}$	2.4×10^{2}	$<4 \times 10^{-2}$
Allyl alcohol	1.9×10^{3}	$<6 \times 10^{-1}$	2.6×10^{3}	$<1 \times 10^{1}$
Acetone	---	---	8.1×10^{2}	$<3 \times 10^{-2}$
Benzene	1.3×10^{2}	$<4 \times 10^{-2}$	1.8×10^{2}	$<1 \times 10^{-2}$

Concentrations reported in ppm (vol/vol).

[a]Ref. 27; product analyzed by gas chromatography--mass spectrometry. [b]Ref. 28; product analyzed by gas chromatography. [c]Not detected by Karl Fischer titration. [d]Lowered to <0.2 ppm by further drying with alumina at -20°C.

the propionitrile content, but this is of little consequence since the properties of propionitrile and acetonitrile are very similar. The protic impurity allyl alcohol may be expected to be generally harmful, while acrylonitrile may interfere in certain reactions (Ref. 31); both impurities and also the added volatile reagents (methanol or octanol) are effectively removed by both procedures. Procedure I, which has some features in common with the procedures previously recommended by this Commission (Ref. 29), appears to have a slight edge over Procedure II. It yields a product which also has excellent voltammetric characteristics; for example, Kiesele (Ref. 28) obtained a current density of less than 0.1 mA cm^{-2} at a platinum disk electrode in 0.2 M Bu_4NBF_4, for a sweep rate of 100 mV sec^{-1} over a potential range of +2.2 to -3.2 V *vs*. an Ag/(10^{-2} M $AgNO_3$ + 10^{-1} M Bu_4NClO_4 in acetonitrile) reference electrode. At least in principle, the pretreatment with octanolate ion and sulfuric acid in Procedure II is drastic; it is not yet totally clear whether or not any new problems are created by this procedure.

DETECTION OF IMPURITIES

The most generally applicable method for the detection and determination of currently common impurities in acetonitrile is gas chromatography, preferably coupled with mass spectrometry (Ref. 27), as illustrated in Table 4. The appropriate experimental conditions for such analyses have been described by Carlsen *et al*. (Ref. 27).

Other methods include Karl Fischer titration of water, ultraviolet spectrophotometric determination of acrylonitrile and benzene, and polarographic determination of acrylonitrile and such potential impurities as ammonia and acetic acid (Ref. 21). The sensitivity of the polarographic determination of acrylonitrile can be enhanced by adding a little water, as is generally the case for compounds requiring protons for their reduction (Ref. 21). "Total acidity" and "total basicity" can be determined by potentiometric titration at the glass electrode with such reagents as tetrabutylammonium hydroxide in 2-propanol and perchloric acid in sulfolane.

REFERENCES

1. D. T. Mowry, *Chem. Revs.* **42**, 189 (1948).
2. M. J. Astle, "*Industrial Organic Nitrogen Compounds*," Reinhold, New York, 1961.
3. M. Walter and L. Ramaley, *Anal. Chem.* **45**, 165 (1973).
4. J. F. Coetzee, in A. Streitwieser, Jr., and R. W. Taft, Eds., "*Progress in Physical Organic Chemistry*," Interscience, New York, Vol. 4, 1967, pp. 45-92.
5. O. Popovych and R. P. T. Tomkins, "*Nonaqueous Solution Chemistry*," Wiley-Interscience, New York, 1981.
6. H. P. Bennetto and E. F. Caldin, *J. Chem. Soc.* A, 2191 (1971).
7. J. A. Riddick and W. B. Bunger, "*Organic Solvents*," Wiley-Interscience, New York, 3rd ed., 1970.
8. C. M. French and D. F. Muggleton, *J. Chem. Soc.*, 2131 (1957).
9. D. S. Berns and R. M. Fuoss, *J. Amer. Chem. Soc.* **83**, 1321 (1961).
10. C. V. Krishnan and H. L. Friedman, in J. F. Coetzee and C. D. Ritchie, Eds., "*Solute-Solvent Interactions*," Marcel Dekker, New York, Vol. 2, 1976, pp. 7-8.
11. A. M. Saum, *J. Polym. Sci.* **42**, 57 (1960).
12. W. Dannhauser and A. F. Flueckinger, *J. Phys. Chem.* **68**, 1814 (1964).
13. M. Michel and E. Lippert, in A. D. Buckingham, E. Lippert, and S. Bratos, Eds., "*Organic Liquids*," Wiley, New York, 1978, pp. 293-308.
14. J. A. Pople and M. Gordon, *J. Amer. Chem. Soc.* **89**, 4253 (1967).
15. Z. Kecki and J. Wojtczak, *Roczniki Chem.* **44**, 847 (1970).
16. J. F. Coetzee and W. R. Sharpe, *J. Solution Chem.* **1**, 77 (1972).
17. I. M. Kolthoff and M. K. Chantooni, Jr., *J. Phys. Chem.* **72**, 2270 (1968).
18. J. F. Coetzee and G. R. Padmanabhan, *J. Phys. Chem.* **69**, 3193 (1965).
19. Y. K. Sze and D. E. Irish, *Can. J. Chem.* **53**, 427 (1975).
20. I. M. Kolthoff and J. F. Coetzee, *J. Amer. Chem. Soc.* **79**, 870 (1957).
21. J. F. Coetzee, G. P. Cunningham, D. K. McGuire, and G. R. Padmanabhan, *Anal. Chem.* **34**, 1139 (1962).
22. G. A. Forcier and J. W. Olver, *Anal. Chem.* **37**, 1447 (1965).
23. J. F. O'Donnell, J. T. Ayres and C. K. Mann, *Anal. Chem.* **37**, 1161 (1965).
24. J. E. McClure and D. L. Maricle, *Anal. Chem.* **39**, 236 (1967).
25. N. S. Moe, *Acta Chem. Scand.* **21**, 1389 (1967).
26. E. O. Sherman, Jr. and D. C. Olson, *Anal. Chem.* **40**, 1174 (1968).
27. L. Carlsen, H. Egsgaard, and J. R. Andersen, *Anal. Chem.* **51**, 1593 (1979).
28. H. Kiesele, *Anal. Chem.* **52**, 2230 (1980).
29. J. F. Coetzee, *Pure Appl. Chem.* **13**, 429 (1966).
30. J. F. Coetzee, unpublished results (1981).
31. American Cyanamid Co., Petrochemicals Dept., "*The Chemistry of Acrylonitrile*," 2nd ed., New York, 1959.

Sulfolane

Prepared for publication by

J. F. Coetzee

INTRODUCTION

Sulfolane (tetrahydrothiophene-1,1-dioxide; tetramethylenesulfone) is one of the more recently introduced members of the class of dipolar aprotic solvents. Some of its physical properties are listed in Table 1. It offers major advantages as a medium for proton transfer and other reactions. It is one of the more inert of the dipolar aprotic solvents with respect to its acid-base properties as well as other types of chemical attack, yet its relative permittivity (dielectric constant) is relatively high. These are generally desirable properties which are to some extent mutually exclusive and which appear to be in optimum compromise in the case of sulfolane.

TABLE 1. Physical properties of sulfolane

Property[a]	Value	Ref.
Freezing temperature, °C	28.45	1
Boiling temperature, °C (1 atm., decomp.)	285	2
Density, g cm^{-3}	1.2625	3
	1.2623	4
Dynamic viscosity, poise	0.1029	4
Refractive index, n_D^{30}	1.4820	5
Dipole moment, μ (in benzene)	4.7	1
Relative permittivity, ε	43.3	3,4

[a] For temperature-dependent properties, temp. = 30°C

The proton basicity of sulfolane is much lower than that of water or even acetonitrile, as shown by its $pK_a(SH^+)$ value of -12.9 (Ref. 6), as compared to ca. -9.5 for acetonitrile (Ref. 7). However, the Lewis base properties of sulfolane are somewhat more pronounced. For example, the donor numbers (with respect to antimony pentachloride) of sulfolane and acetonitrile are similar: 14.8 and 14.1, respectively (Ref. 8), and sulfolane does form an adduct with boron trifluoride (Ref. 9). Sulfolane is a very weak Brønsted acid, as shown by its $pK_a(S)$ value of > 31 (Ref. 10), and it is highly resistant to electrochemical oxidation or reduction (Ref. 3) and to chemical attack, even by the reactive nitronium ion (Ref. 11). Its main limitation for many (but not all) applications is its high viscosity, approximately ten times that of water and thirty times that of acetonitrile.

Typical fundamental studies recently carried out in sulfolane are the following: hydrogen bonding and proton transfer reactions (Ref. 12-15), transport properties of electrolytes (Ref. 4 and 16), and voltammetry of inorganic compounds (Ref. 17). For analytical purposes, the response of hydrogen ion indicator electrodes is of particular interest. In buffer solutions of not too high acidity, the response of both the hydrogen and the glass electrodes is clearly Nernstian; however, there is some doubt whether the same is true for the most acidic solutions ($HClO_4$, $HSbCl_6$) (Ref. 14 and 15).

COMMERCIAL PREPARATION AND KNOWN IMPURITIES

Sulfolane is prepared commercially by Diels-Alder reaction of 1,3-butadiene with sulfur dioxide, giving 2,5-dihydrothiophene-1,1-dioxide (3-sulfolene), which is then hydrogenated in the presence of Raney nickel to give sulfolane (Ref. 18). Principal impurities are reported to be water and 3-sulfolene, while 2-sulfolene and isopropyl sulfolanyl ether may also be present (Ref. 19). The 3-sulfolene present as an impurity dissociates into sulfur dioxide and 1,3-butadiene at temperatures above 70°C, but 2-sulfolene has much greater thermal stability and a boiling temperature of 143°C at a pressure of 5 torr, as compared with 118°C for sulfolane itself. The other impurity, isopropyl sulfolanyl ether, has a boiling temperature of 133-136°C at 5 torr. Finally, in view of the existence of other patents for the preparation of sulfolane (Ref. 20), the possibility exists that pyridine, thiophene and sulfoxides may be present in certain commercial products.

PURIFICATION PROCEDURES

In view of the relative inertness of sulfolane, several of the impurities listed may modify its properties significantly. As for other relatively inert solvents, it is unlikely that an all-purpose purification procedure can be devised which will give optimum results in all possible applications. Consequently, purification should be tailored to the intended use of the solvent.

Several substantially different procedures, differing widely in complexity, have been described. However, in the majority of cases, only limited information has been provided on tests for impurities, so that it is difficult to assess the relative effectiveness of different procedures.

Among the simplest procedures described are those of Morman and Harlow (Ref. 12) and Fernández-Prini and Prue (Ref. 4). The former authors obtained sulfolane suitable for a variety of acid-base titrations by simple treatment of the commercial solvent with activated alumina. The latter workers obtained sulfolane with the low conductivity of 2-5 x 10^{-9} ohm^{-1} cm^{-1}, a water content of less than 5 x 10^{-3} M, and the other properties listed in Table 1, by carrying out two vacuum distillations of a product supplied by Whiffen and Sons, Ltd., the second from solid sodium hydroxide. It appears that this straightforward purification should be suitable for many applications, provided that the starting material is not too contaminated, as has sometimes been our experience.

Benoit *et al.* (Ref. 21) pretreated the solvent with sulfuric acid and hydrogen peroxide (to oxidize possible sulfoxide impurities), extracted it with methylene chloride and, after further pretreatment, distilled it *in vacuo*. This procedure eliminated an optical absorption band at 220 nm and an anodic wave at -1.2 V at the DME *vs.* an Ag/(10^{-2} M Ag^+ in sulfolane) reference electrode observed with the starting material. The product contained 8 x 10^{-2} M water, but solutions of salts could be dried further with molecular sieves or calcium sulfate.

A detailed procedure was developed by Coetzee *et al.* (Ref. 3 and 19) specifically to reduce to a minimum the content of sulfur dioxide and other acidic or basic impurities. A similar procedure was used by Benoit *et al.* in subsequent work (Ref. 13). It has the disadvantages that it is time-consuming and that it leaves some of the (unknown) impurity responsible for the absorption at 220 nm. Commercial sulfolane is heated with 10 g l^{-1} of solid sodium hydroxide at 170-180°C for 24 hours while a slow stream of nitrogen is bubbled through it by means of a gas dispersion tube. This treatment should result in essentially complete thermal decomposition of 3-sulfolene and elimination of sulfur dioxide and other volatile impurities. Sulfolane itself is thermally stable up to 200°C (Ref. 18). The product at this stage has a dark reddish-brown color and a conductivity of *ca.* 5 x 10^{-6} ohm^{-1} cm^{-1}. It is decolorized by stirring with 10 g l^{-1} of decolorizing carbon (e.g., Fisher Norit-A Decolorizing Carbon-Neutral) at 80-90°C for 6 hours, followed by suction filtration (while hot) through a medium-porosity glass filter. The light yellow product is then passed through a 4-foot column packed with a mixed bed of acid "macroreticular" ion exchange resins (Rohm and Haas Amberlyst-15 and Amberlyst-A-21). Conventional ion exchange resins have been shown to function inadequately in many nonaqueous solvents on account of physical breakdown of the resin and unfavorable exchange kinetics; macroreticular resins appear to be superior in these respects (Ref. 22). The column should be heated to 70-80°C to increase the exchange rate and also the flow rate, because the viscosity of sulfolane is high. The effluent should be clear and have a low conductivity of *ca.* 2 x 10^{-9} ohm^{-1} cm^{-1}. Finally, it is vacuum distilled from 5 g l^{-1} of calcium hydride, preferably in a rotary still (e.g., a Büchi rotary vacuum evaporator) at a pressure of 0.01-0.02 torr and an oil bath temperature of 90-100°C. The 80% center cut is collected and transferred under nitrogen to a darkened storage vessel from which it can be dispensed by nitrogen presssure.

TESTS FOR IMPURITIES

Sulfolane purified by the procedure of Coetzee *et al.* had a conductivity of 2×10^{-9} ohm^{-1} cm^{-1}, a water content of 1×10^{-3} M as determined by Karl Fischer titration, and the other properties listed in Table 1. Acidic and basic impurities were determined by potentiometric titration of sulfolane diluted with an equal volume of methanol, using methanolic solutions of sodium methoxide and perchloric acid as titrants and a glass electrode as indicator (limit of detection: *ca.* 1×10^{-6} M, calculated for transfer of one proton). Commercial sulfolane was found to contain 4×10^{-3} and 2×10^{-4} M acidic and basic impurities, respectively, while corresponding concentrations in the purified solvent were 5×10^{-6} and less than 1×10^{-6} M, respectively (Ref. 3 and 19). Voltammetric discharge potentials in 0.1 M Et_4NClO_4 *vs.* an Ag/(0.1 M $AgClO_4$ in sulfolane) reference electrode (AgRE) were the following: +0.04 and -3.50 V at the dropping mercury electrode, and +2.3 and -2.9 V for a current of 2 µA at a platinum electrode of 9-mm^2 area rotated at 60 rad sec^{-1}. (Corresponding potentials *vs.* an aqueous SCE are 0.70 V more positive). Residual currents at the dropping mercury electrode could be considered low, even when making due allowance for the high viscosity of the solvent. For example, at -2.7 V *vs.* AgRE, the residual current (with $m^{2/3}t^{1/6} = 1.0$) was only 0.2 µA, which is of the order expected for the capacitor current alone; the shape of the current-potential trace also indicated that the faradaic component of the residual current was negligible. Sulfur dioxide, which gives cathodic waves at -1.4 and -2.8 V *vs.* AgRE, could not be detected. At the rotated platinum electrode and also at the glassy carbon electrode (Ref. 20), anodic residual currents increased smoothly with increasingly positive potential and revealed no definite waves up to about +2 V which could be attributed to oxidizable impurities. However, the absorption spectrum showed that some of the impurity responsible for the absorption at 200 nm was still present. Furthermore, the solvent had some capacity to decolorize bromine and also alkaline permanganate. Photometric titration with the latter indicated an oxidizable impurity level of *ca.* 3×10^{-4} M, calculated as sulfolene (Ref. 20). If it is important to reduce the concentration of oxidizable impurities to a minimum, it is advantageous to include in the purification procedure a step involving treatment with permanganate. A significant reduction was obtained in absorption at 220 nm and also in the anodic residual current at the rotated glassy carbon electrode (at the expense of a threefold increase in conductivity) by omitting the ion-exchange step in the procedure of Coetzee *et al.* (Ref. 3 and 19) and substituting digestion with potassium permanganate (2 g l^{-1}) for 3 hours at 90°C, followed by filtration (Ref. 20). Alternatively, the procedure of Benoit *et al.* (Ref. 21) may be used.

REFERENCES.

1. L. Jannelli, M. Della Monica, and A. Della Monica, *Gazz. Chim. Ital.* *94*, 552 (1964).
2. R. W. Alder and M. C. Whiting, *J. Chem. Soc.* 4707 (1964).
3. J. F. Coetzee, J. M. Simon, and R. J. Bertozzi, *Anal. Chem.* *41*, 766 (1969).
4. R. Fernández-Prini and J. E. Prue, *Trans. Faraday Soc.* *62*, 1257 (1966).
5. J. W. Vaughn and C. J. Hawkins, *J. Chem. Eng. Data* *9*, 140 (1964).
6. S. K. Hall and E. A. Robinson, *Can. J. Chem.* *42*, 1113 (1964).
7. N. C. Deno and M. J. Wisotsky, *J. Amer. Chem. Soc.* *85*, 1735 (1963).
8. V. Gutmann, "*Coordination Chemistry in Non-Aqueous Solutions*", Springer-Verlag, Wien, New York, 1968, p. 19.
9. J. G. Jones, *Inorg. Chem.* *5*, 1229 (1966).
10. F. G. Bordwell, R. H. Imes, and F. C. Steiner, *J. Amer. Chem. Soc.* *89*, 3905 (1967).
11. J. Jones and J. Jones, *Tetrahedron Letters* *31*, 2117 (1964).
12. D. H. Morman and G. A. Harlow, *Anal. Chem.* *39*, 1869 (1967).
13. R. L. Benoit, A. L. Beauchamp, and R. Domain, *Inorg. Nucl. Chem. Lett.* *7*, 557 (1971).
14. J. F. Coetzee and R. J. Bertozzi, *Anal. Chem.* *45*, 1064 (1973).
15. R. L. Benoit and P. Pichet, *J. Electroanal. Chem.* *43*, 59 (1973).
16. M. Della Monica and L. Senatore, *J. Phys. Chem.* *74*, 205 (1970).
17. J. F. Coetzee and J. M. Simon, *Anal. Chem.* *44*, 1129 (1972).
18. Shell Chemical Co., *Technical Bulletins* IC: 63-13R (Sulfolane) and PD-146 (3-Sulfolene).
19. J. M. Simon, Ph.D. Thesis, University of Pittsburgh, 1969.
20. J. C. Spurgeon, Ph.D. Thesis, University of Pittsburgh, 1972.
21. J. Desbarres, P. Pichet, and R. L. Benoit, *Electrochim. Acta* *13*, 1899 (1968).
22. Rohm and Haas Co., "*Technical Notes on Amberlyst 15* and *Amberlyst A 21*", Philadelphia, Pa. 1968.

Propylene Carbonate

Prepared for publication by

T. Fujinaga and K. Izutsu

PROPERTIES OF PROPYLENE CARBONATE

Propylene carbonate (systematic name 4-methyl-1,3-dioxolan-2-one, sometimes called 4-methyl-dioxolone-2 or propanediol-1,2-carbonate, often abbreviated PC) is an odorless, colorless, aprotic solvent with a wide liquid range, a large dipole moment, and a relatively high relative permittivity (dielectric constant). Several detailed reviews have been published concerning its properties and applications (Ref. 1-3).

Some physical properties of PC are summarized in Table 1.

TABLE 1. Physical properties of propylene carbonate.

Property	Value	Ref.
Freezing temperature,[a] °C	-49.2	4
Boiling temperature, °C		
-at 760 torr	241.7	4
-at 4.5 torr	92	5
Density, g cm^{-3}		
-at 25°C	1.198	6
-at 40°C	1.183	6
Viscosity, cP		
-at 25°C	2.530	6
-at 40°C	1.916	6
Relative permittivity at 25°C	64.92	7
Dipole moment, D	4.94	8
Refractive index, n_D^{20}	1.4209	4
Electric conductivity, ohm^{-1} cm^{-1}		
(lowest value)	1×10^{-8}	1

[a]PC undercools easily; it can remain liquid at temperatures as low as -80°C for periods of several months (Ref. 4).

From dielectric studies, PC is considered to behave as a normal polar liquid with strong dipole-dipole interactions, but with little or no specific association present (Ref. 7 and 10).

PC is not hygroscopic to any great extent, but it gradually takes up moisture if exposed to a humid atmosphere. At 25°C, water and PC are not miscible in all proportions: phase separation occurs between ca. 8 and 82 wt. % of water. However, mutual solubility increases with temperature, and above 61.1°C the two solvents are miscible in all proportions (Ref. 11). PC is miscible in all proportions with a majority of the common organic solvents and it readily dissolves many organic compounds and some inorganic salts (Ref. 1, 4 and 6). It has a high dielectric constant, so that many dissolved ionic compounds are extensively dissociated.

An electron density distribution of the PC molecule has been calculated as shown in Fig. 1 (Ref. 12). Though some evidence has been obtained from infrared and Raman spectroscopic studies (Ref. 13) that PC has a slightly distorted planar ring structure, the CNDO/2 calculation was carried out by assuming a planar ring with the =O bond in the plane. The distribution shows considerable electron density on the ring oxygens as well as on the carbonyl oxygen; it therefore seems that the interaction of cations with PC is not restricted to the

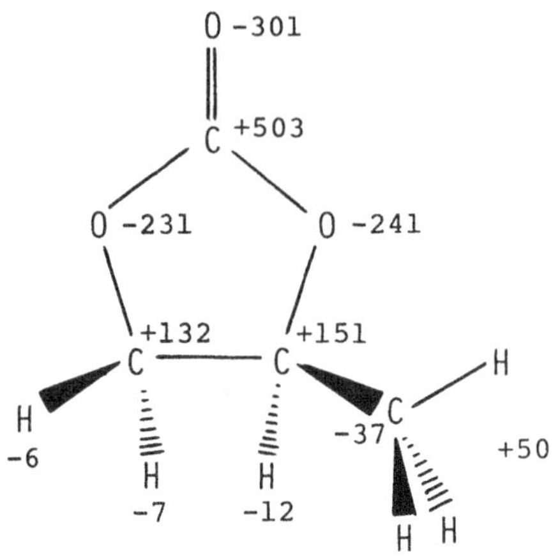

Figure 1. Electron density distribution (in 10^{-3} e.u.) of propylene carbonate (Ref. 12).

carbonyl oxygen alone, and may also involve extensive ring interaction. The basicity of PC is moderately weak: from many results (Ref. 14 and 15) PC seems to be slightly more basic than acetonitrile. The donor number of 15.1 for PC is larger than the value of 14.1 for acetonitrile, but much smaller than the values of 26.6 for dimethylformamide and 29.8 for dimethyl sulfoxide (Ref. 16). The acidity of PC is also weak and approximately the same as those of acetonitrile and dimethyl sulfoxide, as shown by their acceptor numbers (PC 18.3, AN 19.3, and DMSO 19.3) (Ref. 16). Therefore, some similarities are found in the solvation by PC and acetonitrile of both cations and anions, though some cations such as Ag^+ and Cu^+ are specifically and strongly solvated by acetonitrile. The differences in the properties of coordination compounds in these two solvents have been considered to be mainly due to the fact that the PC molecule is large and bulky (Ref. 17).

The autoprotolysis constant of PC has been reported as pK_s = 29.2 (Ref. 18); however, see Chapters 1 and 3. The behavior of Brönsted acids and bases in PC also resembles that in acetonitrile (Ref. 15 and 19). Perchloric acid is almost completely dissociated in PC, but hydrochloric acid is a weak acid (Ref. 19). The pK_a values of picric and salicylic acids are 9.3 and 15.2 (Ref. 15), which are somewhat smaller than the values 11.0 and 16.7 in acetonitrile (Ref. 20), but are much larger than those of ca. -1.0 and 6.8 in dimethyl sulfoxide (Ref. 21) and 1.6 and 8.2 in dimethylformamide (Ref. 22). The phenomenon of homoconjugation between acids and their anions is also observed in PC to approximately the same extent as in acetonitrile (Ref. 15). Ion conductances in PC are relatively small because of its high viscosity (Ref. 1 and 9).

Under ordinary conditions of storage, PC is stable (Ref. 4). Above 200°C, however, appreciable decomposition occurs (Ref. 23 and 24) and traces of acids or bases will promote degradation into propylene oxide, propionaldehyde, allyl alcohol and carbon dioxide (Ref. 4). PC is hydrolysed fairly rapidly in the presence of acids or bases even at room temperature.

PC is fairly stable to oxidation and reduction, and has a wide potential range useful for voltammetric measurements. In PC-0.1 M KPF_6 solution the useful potential range on a platinum electrode extends from +3.2 to -3.6 V vs. Ag/0.1 M Ag^+ (PC) electrode, judging from the potentials at which current densities attain 1 mA/cm^2 (Ref. 25). On a mercury electrode, it ranges from +0.1 to -2.9 V in PC -0.1 M Bu_4NClO_4. PC is relatively stable to attack by halogens anodically liberated at a platinum electrode (Ref. 4). Moreover, normally unstable organic radical ions and alkali metals can be stable in pure PC (Ref. 26 and 27).

PC is essentially non-toxic, either for large single oral doses, by skin absorption, or even on repeated feeding to test animals. While PC is moderately irritating to the mucous membranes of the eyes and respiratory tract, it presents no serious hazards to health in its handling (Ref. 4).

With these convenient properties, PC has been employed widely as a solvent for various synthetic polymers and as a reaction medium (Ref. 4). It has been receiving increasing attention as an electrochemical solvent (Ref. 3 and 6). Its principal use appears to be in the development of aprotic electrolytes for high energy batteries (Ref. 3 and 27). It is also promising as a solvent for ESR studies of free radical species (Ref. 26) and as a solvent for extractions (Ref. 28) and nonaqueous redox titrations (Ref. 29) in analytical chemistry.

TESTS FOR PURITY

Commercially, PC is produced by the reaction of 1,2-propylene oxide with carbon dioxide. The yield from this reaction is low unless a catalyst, such as a quaternary ammonium halide, is used (Ref. 2). The major impurities commonly found in PC therefore include propylene oxide and carbon dioxide, and also 1,2- and 1,3-propanediol (propylene glycol), allyl alcohol, ethylene carbonate, and other unidentified substances.

The following tests can be employed to detect impurities in PC.

(a) Water content by Karl Fischer titration. The coulometric method is preferable because the sample volume needed is smaller than in the conventional electrometric titration method.

(b) Gas chromatographic measurements. This method is most useful. Jasinski and Kirkland (Ref. 24) developed a method using a column of Poropak Q for the determination of water and volatile organic impurities. The analytical conditions recommended by them are shown in Table 2.

TABLE 2. Conditions for analysis of propylene carbonate.

Column	2 ft Poropak Q, presieved (80-100 mesh)
Carrier gas-thermal conductivity cell	Dry helium, 100 ml/min
Detector temperature	200°C
Filament current	200 mA
Carrier gas-flame ionization detector	Argon + 1% hydrogen, 180 ml/min
Sample size	9 to 40 μl
Column temperature	70°C for analysis of water
	110°C for analysis of low boiling organics
	140°C for analysis of high boiling organics
Injector temperature	Variable (below 140°C)

From Ref. 24.

At a column temperature of 70°C, carbon dioxide and water can be detected by using thermal conductivity cells. Propylene oxide, allyl alcohol and some other impurities are detected at a column temperature of 110°C, and 1,2- and 1,3-propanediols with a column at 140°C. A flame ionization detector is used for these purposes. The components are identified by matching the retention times with those of pure materials. Determinations can be carried out using standards made by the addition of the pure components to the purified PC. For a 40-μl sample, the limits of detection are 1.5 ppm for water, 0.05 ppm for propylene oxide, and 0.5 ppm for propanediol. Satisfactory results have also been obtained using different column conditions (Ref. 30 and 31). Denat et al. (Ref. 30) could best separate the peak of ethylene carbonate from that of PC by using a column of 10% Carbowax 20 M and Chromosorb W (70 - 80 mesh).

(c) Polarographic measurements. When tetraethylammonium perchlorate is used as the supporting electrolyte, a deaerated solution prepared from moderately pure PC has a polarographic range extending from +0.5 to -2.5 V vs. aq. SCE at a mercury electrode and from +1.7 to -1.9 V vs. aq. SCE at a platinum electrode (Ref. 26). (The useful potential range is influenced significantly, especially on a platinum electrode, by the presence of traces of water and by the kind of supporting electrolyte. For details, see Ref. 25). Propanediol impurity shows an anodic wave at a platinum electrode polarized in the positive direction. The uncharged, basic impurity which is detectable by conductometric titration with a weak acid (see below) can give a wave at around -2.0 V vs. SCE at a mercury electrode. Water cannot be detected at a mercury electrode, but, in chronopotentiometry or cyclic voltammetry at a platinum electrode in $LiClO_4$-PC solutions, water from 20 to 500 ppm has been reported to give cathodic and anodic waves between +2.0 and +1.3 V vs. Li/1 M Li^+ electrode (Ref. 27 and 32).

(d) Conductivity measurements. It is relatively easy to prepare PC that has an electrolytic conductivity of 1 - 5 x 10^{-8} ohm^{-1} cm^{-1}, but this method is only useful in the detection of ionic impurities. Conductometric titrations of PC with a solution of p-toluenesulfonic acid (a weak acid in PC) is sometimes effective for the detection and determination of an uncharged, basic impurity which is usually found in PC as received (Ref. 15). For the elimination of this impurity, see below.

(e) Spectrophotometric measurements. Pure PC shows little or no absorption in the UV region between 240 and 340 nm, but impure PC does. Therefore, the measurement of the UV spectrum in this region seems to be a convenient test for the purity of PC, though the impurities that give the absorption have not yet been identified.

Water at and above the 20 ppm level in PC and in its solution with $LiClO_4$ can be determined by differential spectrophotometry employing the 1.9 μm near-infrared absorption band (Ref. 33).

PURIFICATION OF PROPYLENE CARBONATE

PC can be purified fairly easily by fractional distillation under reduced pressure. However, because the presence of trace impurities shows marked effects on the performance of this solvent (see Chapter 1), detailed studies have been carried out for the preparation of highly pure solvent. Of these, the study of Jasinski and Kirkland (Ref. 24) is most often referred to. They provide the following thorough description of their procedure.

"Before distillation, 3500 ml of the commercial PC (practical grade 120-122°C/17 mm Hg from Matheson, Coleman and Bell) is percolated at a rate of 10 ml/min down a column of 220 g of dried molecular sieves (Linde 5A, dried by heating at 350°C for 14 hours under a stream of dry argon). By this procedure, the water impurity is reduced to below the limit of detection (<2 ppm).

Distillation is carried out with a Podbielniak Semi-Cal (Series 3650) adiabatic vacuum fractional distillation column. The 'tilting funnel' system is used to control the reflux ratio; the 2 x 72 in column is filled with Heli-pak (Nichrome). The total charge of PC in the distillation kettle is 3500 ml. A vacuum of 0.5 to 1 mm Hg is achieved at the kettle and its temperature is maintained at 110° to 111°C. The column head temperature levels off between 72° and 75° after the first 100 ml of distillate is collected. A boil-up rate of 200 ml/h is used and the reflux ratio is held between 10:1 and 20:1. The drip spout is placed in the receiver arm to prevent contact of the distillate and the joint. Grease is avoided; Teflon sleeves with Hypalon O-rings are used at the joint; Teflon stopcocks are used to isolate sections of the receiver arm. Distillates are collected and immediately transferred to the inert atmosphere glove box. Table 3 shows the results of gas chromatographic tests for impurities in each section of the distillates.

For further purification, the final 1500 ml of the distillation run described in Table 3 are combined and the low boiling materials are evaporated by passing dry nitrogen gas through the solvent for five days. By this evaporation procedure, lowboiling organics are reduced to below 0.4 ppm and water becomes undetectable (below 1.5 ppm); the glycol concentration, however, does not change. The solvent is then redistilled; small amounts of low boiling components (0.7 ppm in total) are detectable only in the first fractions of the distillate and the content of glycol levels off at approximately 11 ppm."

TABLE 3. Distillation run.

Cumulative volume, ml	Concentrations, ppm			
	Water	Propylene oxide	Allyl alcohol	Propane-diol
Starting material	450	33	10	2000
72	35	400	--	--
302	<2	24	--	820
406	<2	6	--	220
856	12	12	2	95
959	9	9	--	--
1159	<2	10	2	58
1266	<2	5	--	49
1466	<2	9	--	27
1601	<2	9	2	15
1801	9	9	3	17
1966	9	6	1	9
2156	4	8	--	12

From Ref. 24.

In many electrochemical investigations, procedures which are modifications or simplifications of the above have been employed for the purification of PC. The following procedure can be recommended for general purposes.

Commercial PC is first dried by keeping it overnight with molecular sieves or by passing it through a column of dried sieves (Linde 4A or 5A). Then the PC is subjected to fractional distillation two or three times under vacuum (one to several torr); the middle 2/3-fraction of the distillate is collected in each distillation and immediately transferred to an inert atmosphere dry box. The use of an adiabatic vacuum distillation column is highly desirable. By this procedure, the contents of water and organic impurities can generally be reduced to below 20 ppm.

Activated alumina (Ref. 30 and 34) and dried calcium oxide (Ref. 9 and 24) have also been used as drying materials before distillation. Distillation from calcium oxide has been reported to reduce the content of propanediols to about 20 ppm but to increase the content of propylene oxide above its level in the starting material itself (Ref. 24). Coetzee and Umemoto (Ref. 31) compared the results of distillation from anhydrous calcium sulfate with that without calcium sulfate. After treatment with molecular sieves and two preliminary distillations from anhydrous calcium sulfate under nitrogen at a pressure of 1 torr through a 25-in. vacuum-jacketed column packed with glass helices, they carried out a third distillation in three different ways: (A) as for the preliminary distillations, at a distillation rate of 60 ml/h; (B) as for (A), but with the calcium sulfate omitted; (C) as for (B), but in a Perkin-Elmer Model 251 Auto Annular Still with a Teflon spinning band, at a reflux ratio of 5:1 to 10:1 and a distillation rate of 10 ml/h. The middle 2/3-fraction of the solvent was collected. The results of the impurity analyses are shown in Table 4. The final

TABLE 4. Impurity levels (mM) in propylene carbonate obtained by different methods.

	Method[a]		
	A	B	C
Water	2 - 3	4	4
1,2-Propanediol	0.7	0.03	0.03
Allyl alcohol	0.02	Trace	Absent
Propylene oxide	0.02	Trace	Absent

[a]For methods A, B, and C, see text.

From Ref. 31.

distillation from calcium sulfate (method A) has the disadvantage that it produces solvent containing significant amounts of propanediol without offering any offsetting advantage of reducing the water content significantly.

PC from some commercial sources shows serious UV absorption due to an unidentified impurity, and it is difficult to eliminate the impurity by the conventional distillation method. Most simply, it is recommended to use PC from other sources which is free from the impurity. However, Denat et al. (Ref. 30) proposed a method to eliminate the impurity by treatment with 10 g/Kg of potassium permanganate. The MnO_2 formed was eliminated by filtration, and the excess $KMnO_4$ dissolving was destroyed by warming at 120°C. (Caution: Potential explosion hazard!) The solution was then clarified by rapid distillation through a Vigreux column. The solution, after being passed through a column of activated alumina, was carefully distilled under vacuum. By this method, the impurity showing the UV absorption between 240 and 340 nm was almost completely eliminated . The resistivity of the middle fraction was 6 to 8 x 10^8 ohm cm. The water content was less than 20 ppm. Talarmin et al. (Ref. 19) also eliminated the same impurity by distillation from $KMnO_4$.

Traces of the basic impurity which can be detected by conductometric titration (see below) interfere with acid-base equilibrium measurements. To eliminate this impurity effectively, Izutsu et al. (Ref. 15) distilled PC, after treatment with molecular sieves, two times from p-toluenesulfonic acid added in about twice the equivalent amount of the impurity. They carried out a third distillation without additives.

PC after distillation is often stored over molecular sieves (usually Linde 4A) under an inert gas atmosphere. This treatment is useful for removing traces of water and also impurities which react with silver ion (Ref. 35), but very small amounts of ionic substances may be introduced, as detected by measurement of electrolytic conductivity.

Electrolyte solutions prepared from pure PC are also dried by treatment with molecular sieves. The process is usually effective, but care should be taken when molecular sieves are attacked by electrolytes. PC-Et_4NF solution has been reported to attack molecular sieves (Ref. 36). Jansta *et al.* (Ref. 37) prepared highly dried PC-electrolyte solution using Na-K alloys and vacuum line technique. In a glass assembly under vacuum, they pre-dried PC with molecular sieves, prepared 0.5 M KPF_6 solution, dried it with Na-K alloy, and carried out electrochemical measurements. The treatment was effective to yield solutions free from such active substances as water and propanediol, but contamination occurred with traces of alkali metal glycolate, potassium carbonate, and an unidentified colored product of PC decomposition.

REFERENCES

1. W. H. Lee, "Cyclic Carbonates," in J. J. Lagowski, ed., "The Chemistry of Non-Aqueous Solvents," Vol. IV, Academic Press, New York, 1976, Ch. 6.
2. L. M. Mukherjee, CRC Crit. Rev. Anal. Chem. 5, 325 (1975).
3. R. J. Jasinski, Advan. Electrochem. Electrochem. Eng. 8 253-335 (1971).
4. Jefferson Chemical Co., "Propylene Carbonate," Technical Bulletin, 1960.
5. R. L. Kronick and R. M. Fuoss, J. Amer. Chem. Soc. 77, 6114 (1955).
6. W. Harris, "Electrochemistry in Cyclic Esters," Ph.D. Thesis, University of California, Berkeley (1958).
7. R. Payne and I. E. Theodorou, J. Phys. Chem. 76, 2892 (1972).
8. R. F. Kempa and W. H. Lee, J. Chem. Soc., 1936 (1958).
9. L. M. Mukherjee and D. P. Boden, J. Phys. Chem. 73, 3965 (1969).
10. L. Simeral and R. L. Amey, J. Phys. Chem. 74, 1443 (1970).
11. N. F. Catherall and A. G. Williamson, J. Chem. Eng. Data 16, 335 (1971).
12. H. L. Yeager, J. D. Fedyk, and R. J. Parker, J. Phys. Chem. 77, 2407 (1973).
13. R. A. Pathrick and A. D. Wilson, Spectrochim. Acta 30A, 1073 (1974).
14. R. L. Benoit and C. Louis, "Solvent Basicity," in J. J. Lagowski, ed., "The Chemistry of Non-Aqueous Solvents," Vol. VA, Academic Press, New York, 1978, Ch. 2.
15. K. Izutsu, I. M. Kolthoff, T. Fujinaga, M. Hattori, and M. K. Chantooni, Jr., Anal. Chem. 49, 503 (1977); K. Izutsu, T. Nakamura, and I. Iijima, to be published (1979).
16. V. Gutmann, Electrochim. Acta 21, 661 (1976).
17. V. Gutmann, "Coordination Chemistry in Non-Aqueous Solutions," Springer, Vienna, 1968, p. 142.
18. N. Sh. Aldarova, K. E. Batlaev, B. B. Tanganov, A. A. Izyneev, and N. A. Baranov, Zh. Fiz. Khim. 50, 759 (1976).
19. J. Talarmin, M. L'Her, and J. Courtot-Coupez, J. Chem. Res., (M) 0327 (1977); ibid., (S) 28 (1977).
20. I. M. Kolthoff and M. K. Chantooni, Jr., J. Amer. Chem. Soc. 87, 4428 (1965); J. Phys. Chem. 70, 856 (1966).
21. I. M. Kolthoff, M. K. Chantooni, Jr., and S. Bhowmik, J. Amer. Chem. Soc. 90, 23 (1968).
22. I. M. Kolthoff, M. K. Chantooni, Jr., and H. Smagowski, Anal. Chem. 42, 1622 (1970).
23. C. E. Levin and A. L. Shapiro, Khim. Prom. 41, 264 (1965).
24. R. J. Jasinski and S. Kirkland, Anal. Chem. 39, 1663 (1967).
25. J. Courtot-Coupez and M. L'Her, Bull. Soc. Chim. Fr., 1631 (1970).
26. R. F. Nelson and R. N. Adams, J. Electroanal. Chem. 13, 184 (1967).
27. R. J. Jasinski, Electrochem. Technology 6, 28 (1968).
28. B. G. Stephens and H. A. Suddeth, Anal. Chem. 39, 1478 (1967); B. G. Stephens and H. L. Felkel, Jr., ibid. 47, 1676 (1975).
29. R. D. Krause and B. Kratochvil, Anal. Chem. 45, 844 (1973).
30. A. Denat, B. Gosse, and J.-P. Gosse, J. Chim. Phys. 72, 343 (1975).
31. J. F. Coetzee and K. Umemoto, Inorg. Chem. 15, 3109 (1976).
32. B. Burrows and S. Kirkland, J. Electrochem. Soc. 115, 1164 (1968).
33. R. Jasinski and S. Carroll, Anal. Chem. 40, 1908 (1968).
34. I. Piljac and R. T. Iwamoto, J. Electroanal. Chem. 23, 484 (1969).
35. J. Courtot-Coupez and M. L'Her, Bull. Soc. Chim. Fr., 675 (1969).
36. J. C. Synnott, D. R. Cogley, and J. N. Butler, Anal. Chem. 44, 2247 (1972).
37. J. Jansta, F. P. Dousek, and J. Riha, J. Electroanal. Chem. 38, 445 (1972).

Dimethyl Sulfoxide

Prepared for publication by

C. G. Karakatsanis and T. B. Reddy

INTRODUCTION

Dimethyl sulfoxide (DMSO) is one of the most prominent members of the class of dipolar aprotic solvents. Members of this class are characterized by very weak hydrogen bond donor properties, high dipole moments and relatively high relative permittivities (dielectric constants); some members, such as DMSO, also possess marked hydrogen bond acceptor properties. Considerable interest in DMSO has developed over the past two decades, primarily because of its versatility, both as a medium of wide applicability for a variety of organic and inorganic reactions and as a reagent with some exceptional biological properties. An excelllent monograph on the chemistry of DMSO with an extensive list of references, reviewing the subject up to 1969, has been published by Szmant (Ref.1).

The versatility of DMSO is mainly associated with the fact that it can act as a nucleophile at either the oxygen or the sulfur atom of the sulfoxide group, thus behaving as either a hard or a soft base, respectively. Its basic properties were recognized as early as 1866 at the time of its discovery by Saytzeff, who isolated a number of DMSO salts of mineral acids (Ref. 1). Almost a hundred years later, the structure of such a crystalline DMSO adduct with nitric acid was studied by infrared spectroscopy (Ref. 2). The proton basicity of DMSO is greater than that of water by some 1.5 ± 0.5 pK units (Ref. 3). DMSO, however, is a very weak Brønsted acid as shown by its $pK_a(S)$ value of ca. 33 (Ref. 1 and 4). The autoprotolysis constant is estimated to have a value of ca. 10^{-32} M^2 (Ref. 4). This is some fifteen powers of ten smaller than a previous estimate (Ref. 5) and as Ritchie points out (Ref. 4), the discrepancy demonstrates again the potentially large errors that can result from trace impurities in dipolar aprotic solvents. (Also see Chapter 1). Thus, it is probably wise to consider even the newer estimate only as an upper limit.

Examples of typical studies that have been carried out in DMSO include the following: determination of pK_a values, both potentiometrically using a glass electrode (Ref. 4-7) and by the indicator overlap technique (Ref. 4, 8, and 9); investigations of intramolecular hydrogen bonding in acid-base equilibria (Ref. 10 and 11); determination of enthalpies of transfer of various ions and molecules from water (Ref. 12); polarographic and, in general, electrochemical studies of a variety of systems (Ref. 13-16); and mechanistic investigations of organic (Ref. 17) and ligand-substitution reactions (Ref. 18). Recently, a thermochemical method for direct comparision of acidities of many types of compounds with a wide range of pK_a-values in DMSO has been described by Arnett et al. (Ref. 19).

As mentioned above, the glass electrode has been used in anhydrous DMSO with considerable success. Butler (Ref. 16) reviewed its use in this solvent as well as in other aprotic media. It was noted that considerable time is needed for glass electrodes to reach equilibrium in certain aprotic solvents; in the case of DMSO, as much as 45 minutes is required. In addition to conventional glass electrodes with aqueous internal solutions, modified electrodes with aprotic internal solutions have also been used. Ritchie and Uschold (Ref. 6), for example, replaced the internal solution of a commercial electrode with a solution of 0.02 M silver perchlorate and 0.006 M p-toluenesulfonic acid in DMSO, with considerable improvement in response. The latter electrode was found to be reversible with respect to hydrogen ion activity over a range of 25 pH units.

PHYSICAL PROPERTIES, PARAMETERS AND STRUCTURE OF DMSO

DMSO is a colorless, odorless, very hygroscopic liquid with a slightly bitter taste and sweet after-taste. Douglas (Ref. 22) determined its vapor pressure in the temperature range from 20 to 50°C by a gas-saturation method. The results obtained could be expressed by

$$\log P = 26.49558 - 3539.32/T - 6.00000 \log T$$

where P is given in torr and T in degrees Kelvin. It was estimated (Ref. 22) that the above equation gives vapor pressure values, in the temperature range of the measurements, to within

$\pm 1\%$. Additional physical properties of DMSO are listed in Table 1 and selected chemical parameters are given in Table 2.

TABLE 1. Selected physical properties of dimethyl sulfoxide.

Property[a]	Value	Ref.
Formula weight	78.134 $\pm$ 0.003	20
Freezing temperature, °C	18.55 $\pm$ 0.02	21
Boiling temperature (1 atm.), °C	189.0 $\pm$ 0.2	21
Vapor pressure, torr	0.600	22,23
Viscosity, cP	1.996	24
Relative permittivity	46.7	25
Refractive index, n_D^{25}	1.4773	26,24
Dipole moment, D	4.11	27
	3.9	28
Density, g cm^{-3}	1.09575	26
Enthalpy of vaporization, kcal mol^{-1}		
at 25°C	12.64	22
at normal b.t.	10.31	24
	10.7	29
Entropy of vaporization, cal mol^{-1} K^{-1}		
at normal b.t.	23.16	29
Flash temperature, open vessel, °C	95	24
Surface tension, dyne cm^{-1}	42.86	30
Specific conductance (20°C), ohm^{-1} cm^{-1}	3×10^{-8}	21
(lowest value measured)	2×10^{-8}	5,14

[a] For temperature-dependent properties, temperature is 25°C except where noted otherwise. For a compilation of some of these properties at other temperatures as well, see Ref. 24 and 31.

TABLE 2. Some chemical parameters of dimethyl sulfoxide.

Parameter[a]	Value	Ref.
Polarizability, A^3 $molecule^{-1}$	7.97	21
Kirkwood g-factor[b]	1.07	31
Kosower Z-value[c]	71.1	32
Dimroth E_T-value[d]	45.0	32
Gutmann donor number[e]	29.8	33
Autoprotolysis constant, M^2 (maximum value)	ca. 10^{-32}	4

[a] Temperature is 25°C.

[b] A measure of the degree of self-association of the solvent, evaluated from dielectric properties.

[c] Solvent polarity parameter, based on the longest wavelength charge-transfer absorption band of 1-ethyl-4-carbomethoxypyridinium iodide.

[d] Solvent polarity parameter, based on the wavelength of the intramolecular charge-transfer band of pyridinium phenol betaine.

[e] A measure of the electron-donor strength of the solvent, defined as the enthalpy (as $-\Delta H°$, in kcal mol^{-1}) for the reaction of the solvent with antimony pentachloride in 1,2-dichloroethane.

In his review, Szmant (Ref. 1) has presented a detailed account of the structural characteristics of DMSO both in the solid and liquid phase. The molecule can be visualized as a tetrahedron, with the sulfur atom located at the center and two methyl groups, an oxygen atom and a nonbonding electron pair occupying the apex positions. The high values of the enthalpy and entropy of vaporization (Table 1) indicate a highly associated liquid. Various studies of DMSO in benzene and carbon tetrachloride support this view of extensive self-association and, furthermore, indicate a strong concentration dependence of the nature of the aggregates formed. An infrared study of DMSO solutions in carbon tetrachloride at 27°C (Ref. 1) is consistent with the existence of essentially monomeric species at low DMSO concentrations (up to 0.008 M) while at higher concentrations (up to 0.3 M), dimers involving the S-O bonds, possibly in a cyclic configuration, are believed to predominate. The existence of cyclic dimers is also suggested by analysis of relative permittivity data for solutions of DMSO in benzene at 25°C (Ref. 27), although the dimeric aggregate in this case has its maximum concentration at a higher DMSO content. In both benzene and carbon tetrachloride at still higher DMSO concentrations, further aggregation, possibly into chainlike form, is believed to occur. For pure, liquid DMSO at moderate temperatures, a chainlike structure formed by alignment of S-O dipoles is favored (Ref. 1). This view has been challenged (Ref. 34), however, on the basis of the Kirkwood g-factor being close to unity, in favor of strong but non-specific dipole-dipole interactions. Whatever the structure of the pure liquid is, however, it appears to suffer some breakdown between 40 and 60°C, since the temperature coefficients of viscosity, density, and refractive index exhibit considerable changes in this temperature range (Ref. 21).

COMMERCIAL PREPARATION AND KNOWN IMPURITIES

The DMSO which is commercially available in the U.S.A. (Crown Zellerbach Corporation, Chemical Products Division, Camas, Washington) is synthesized from dimethyl sulfide, a by-product of the manufacture of paper. The presence of an unidentified organic impurity is shown by the NMR spectrum of the unpurified solvent (Ref. 35). One European product (Union Rheinische Braunkohlen Kraftstoff A. G., Wesseling, Germany) is reported to contain 0.4% water, a small amount of dimethyl sulfide, and a trace of dimethyl sulfone (Ref. 21). As mentioned in the previous section, DMSO is highly hygroscopic with an equilibrium moisture content of ca. 10% when in contact with air at 20°C (Ref. 24). Thus, the solvent will readily absorb water vapor if exposed to the atmosphere. DMSO of the highest commercial purity contains very little impurity besides water (Ref. 16). In fact, water was the only electroactive or volatile impurity found in commercially available spectroscopic grade DMSO (Ref. 36). Any organic impurities present are most likely limited to decomposition products which include, in addition to dimethyl sulfide and dimethyl sulfone, methyl mercaptan and bis(methylthio)-methane; the latter two compounds are products of acid-catalyzed decomposition of DMSO (Ref. 16).

PURIFICATION PROCEDURES

Sears, Lester and Dawson (Ref. 37) used the following procedure. DMSO (99.9%) was refluxed over calcium oxide for several hours and then fractionated through a column at a pressure of about 5 torr. The middle fraction was collected and fractionated again one or more times until the specific conductance was 3 x 10^{-8} ohm^{-1} cm^{-1} or less. The density, viscosity and relative permittivity were all measured and compared with known values. The water content of the solvent, however, was not determined. This procedure was criticized (Ref. 14) because refluxing with calcium oxide (or barium oxide) causes extensive decomposition of the solvent and necessitates removal of the decomposition products. Furthermore, only about 30% of the water is removed in each distillation cycle when calcium oxide is used (Ref. 38) and a more efficient dehydrating agent is needed.

Gutmann and Schober (Ref. 13) subsequently used the following method. Technical grade DMSO was triply distilled from itself through a 50-cm packed fractionating column at a pressure of 12 torr. Distillation over phosphorus pentoxide or sodium metal led to decomposition of the solvent starting at 60°C and was not recommended. The fraction boiling at 81°C was collected and the residue discarded. The melting temperature and the specific conductance at 21°C of the purified solvent were 18°C and 4.5 x 10^{-7} ohm^{-1} cm^{-1}, respectively. The authors felt that this method is adequate for polarographic use since no change in the polarographic waves of a number of metal ions was observed when the solution was exposed to the atmosphere for an hour. Nevertheless, the conductivity indicates that this method does not reduce the water content effectively and that an efficient dehydrating agent which does not decompose the solvent is needed.

Schlafer and Schaffernicht (Ref. 21) have desribed an elaborate fractionation system to produce pure DMSO. The first 20% fraction is distilled into one part of the system to remove water and dimethyl sulfide. This part of the system is then shut off and the product, about 70% of the starting material, collected in another part of the apparatus. Using a 1-m long column of 5-cm internal diameter, packed with glass Raschig rings, solvent containing only 4 ppm of water was obtained. The specific conductance of this solvent, 3 x 10^{-8} ohm^{-1} cm^{-1}

at 20°C, is no better than that obtained by other methods, however. Schlafer and Schaffernicht advocated using Teflon sleeves lubricated with DMSO to seal ground glass joints because of the solubility of silicone greases in DMSO.

Kolthoff and Reddy (Ref. 5, 14 and 38) recommended the following purification process. DMSO was shaken overnight with acidic activated alumina (Woelm chromatographic grade-activity I) which had been fired to 600°C and cooled. The solvent was then siphoned into a distillation flask and vacuum fractionated under a nitrogen atmosphere at a pressure of a few torr through a heated 20-in column packed with porcelain saddles. Some alumina was transferred in the siphoning operation. The first fraction, amounting to 10% of the total, was rejected and the second fraction of about 80% of the starting material was collected. The procedure was repeated two more times or until the specific conductance was 2 x 10^{-8} $ohm^{-1}cm^{-1}$ or less and the water content was 0.01% by weight or less as determined by the Karl Fischer method. This procedure removes about 65% of the water present for each cycle of purification and has the advantage that a significant amount of water is removed through the use of a mild dehydrating agent without decomposition of the solvent. Basic impurities should also be removed by the acidic alumina. The use of a more efficient dehydrating agent would reduce the number of cycles necessary to obtain anhydrous solvent.

A similar procedure was employed by Menashi, Reynolds and Van Auken (Ref. 39). DMSO was first fractionally distilled at a pressure of 2-3 torr at about 50°C. The middle two-thirds of the distillate was collected and shaken overnight with alumina. It was then fractionally distilled under the same conditions as before and the middle one-half was collected for use. The water content was found to be approximately 0.015 M. The authors used the solvent purified in this manner for iron(II) and iron(III) isotope exchange studies. They note that both the commercial material as obtained from the manufacturer (Crown Zellerbach Corp.) and a 99.9% analyzed material (J. T. Baker) gave identical results after purification.

Price and Whiting (Ref. 40) employed the following procedure. DMSO was purified by adding a small amount of triphenylmethane indicator to the commercial material and then dimsylsodium solution until a strong red color developed. Dimsylsodium solution, a solution of the sodium salt of the lyate ion of DMSO (methylsulfinylcarbanion) in DMSO, was prepared by the method of Corey and Chaykovsky (Ref. 41 and 42). DMSO (500-600 ml) was reacted with an equimolar amount of sodium hydride at 65-70°C under a nitrogen atmosphere for 45 minutes.

$$NaH + (CH_3)_2SO \rightarrow H_2\uparrow + Na^+ + CH_3(SO)CH_2^-$$

Following treatment with dimsylsodium solution, the solvent was distilled at 0.5 torr and 40°C. Dimsylsodium reacts with water and any other substances that are stronger acids than DMSO.

$$Na^+ + CH_3(SO)CH_2^- + H_2O \rightarrow NaOH\downarrow + (CH_3)_2SO$$

$$Na^+ + CH_3(SO)CH_2^- + HX \rightarrow Na^+ + X^- + (CH_3)_2SO$$

The products are either insoluble or ionic and will not distill. The triphenylmethane acts as an indicator since excess dimsylsodium reacts to produce the triphenylmethyl anion with a resulting red color. The disadvantages of the method are that considerable effort is involved in the preparation of the dimsylsodium solution and the fact that basic impurities are not removed by the treatment. The authors used the solvent purified in this manner for titrimetric purposes with considerable success.

Maricle and Hodgson (Ref. 43) employed the following purification procedure. DMSO was heated over sodium hydroxide at 90°C for two hours and then flash-distilled under vacuum. The method is apparently very efficient since the water content, as determined by the Karl Fischer method, is reduced to 0.01% by a single treatment; however, no analysis for organic impurities was reported.

Bovey, Longworth and Flood (Ref. 35) have reported that the NMR spectrum of unpurified DMSO contains lines due to two impurities, one of which is water. Treatment with calcium hydride in a nitrogen atmosphere eliminates both lines. Roberts and Sawyer (Ref. 44) employed purified commercial grade DMSO which was vacuum-distillled over calcium hydride (4-40 mesh, 10 g/ℓ) for voltammetric determination of dissolved carbon dioxide. The authors note that the residual currents observed in DMSO were about twenty times smaller than those in acetonitrile which had been purified by successive distillation from phosphorus pentoxide and calcium hydride. More recently, Coetzee _et al._ (Ref. 18) employed a similar procedure to purify DMSO used as solvent in kinetic studies of ligand-substitution reactions of nickel(II) complexes. Reagent grade DMSO was vacuum-distilled, first alone and then from calcium hydride. The water content of the product, as determined by Karl Fischer titration, was _ca._ 0.004 M.

Ritchie et al. recommended the following purification procedure originally proposed by Steiner (Ref. 45 and 46). DMSO was dried over molecular sieves, distilled under vacuum, and finally redistilled from freshly prepared potassium amide, also under vacuum, in a rotary evaporator. The distillation temperature was kept well below 50°C and the distillate was held under vacuum for several hours to remove traces of ammonia. The purified product was reported to contain less than 5 x 10^{-6} M acidic or basic impurities and less than 10 ppm water (Karl Fischer titration).

Mann (Ref. 47) has recommended purification by a procedure involving fractional crystallization. About 80% of a solvent batch was slowly frozen. The liquid was then discarded and the solid melted. Fifty ml of benzene per liter of solvent was then added and the mixture was distilled to remove water. The remaining DMSO was very rapidly distilled under vacuum. The product contained some water but only a small amount of other low-boiling impurities. Mann states that a drier product could be obtained by using 5A molecular sieves before distillation. Controlled crystallization has also been used by Molochko and Kurdyumov (Ref. 48). The authors employed forced convective mixing of the liquid phase (in a vertical column apparatus) in the crystallization process. For DMSO purification at a crystallization rate of 0.8 cm per hour, the residual impurities content, as determined by gas chromatographic analysis, showed a signficant decrease by this method as compared to crystallization purification without mixing.

Butler and Cogley (Ref. 49 and 50) have used Linde 5A molecular sieves to remove water efficiently from DMSO. The sieves were first dried in argon for 16 hours at 500°C. Matheson, Coleman and Bell chromatographic or spectrograde DMSO was then treated with the cooled sieves for a period of one week, following which the solvent was filtered through a 10-20 micron glass frit to remove suspended matter and then used directly. The product was found to contain less than 10 ppm of water and less than 50 ppm of low-boiling organic impurities (Ref. 49). Butler and Cogley (Ref. 50) have also shown that treatment of nonaqueous solutions of hydrated salts with molecular sieves can significantly reduce the water content. This technique can be particularly useful for solutions of salts that are difficult to dry, such as tetraalkylammonium compounds. Butler and Cogley have also found that some ion exchange occurs between the electrolyte and the sieves introducing small amounts of sodium ion into the solution.

RECOMMENDED PROCEDURE

Although treatment with pre-dried type 5A molecular sieves reduces the impurity level remarkably (Ref. 49 and 50), it may also introduce some ionic impurities into the solvent. Careful vacuum distillation from calcium hydride after filtration serves to remove ionic as well as other impurities and any residual suspended matter from the sieves.

TESTS FOR PURITY

(A) Karl Fischer determination of water.
The Karl Fisher method may be used for the determination of water in DMSO. Especially for samples containing less than 100 ppm of water, coulometric generation of Karl Fischer reagent with electrometric equivalence-point detection can be employed with advantage (Ref. 51). Commercially available coulometers, such as the Photovolt Aquatest II model, which employ coulometric generation of iodine and a direct digital readout of the water content, are very convenient for this purpose. Classical titrations may be performed as follows (Ref. 38). Ten ml of reagent-grade methanol is introduced into a 25-ml volumetric flask and titrated with Karl Fischer reagent to a definite brown color due to excess iodine. Then, 1 ml of DMSO is added, and the solution is again titrated to the same color. The Karl Fischer reagent is standardized in the same manner, except that water is introduced from an ultramicroburet instead of the sample DMSO.

(B) Gas chromatographic determination of water and organic impurities.
Cogley and Butler (Ref. 49, 50 and 16) have employed gas chromatography to determine both water and low-boiling organic impurities. They recommend the use of Poropak Q, a polystyrene packing material, according to the procedures developed by Jasinski and Kirkland (Ref. 52) for the determination of impurities in propylene carbonate. Severe "tailing" of the water peak was observed with other adsorbents and Cogley and Butler emphasize the need for careful control of inlet temperature and other variables in order to obtain reliable gas chromatographic data. The sensitivity of the method is stated to be less than 2 ppm of water in the solvent (Ref. 16).

(C) NMR
Water is detectable in DMSO by an NMR line at δ = 3.25 vs. tetramethylsilane (Ref. 35). The unknown organic impurity gave a line at δ = 2.0. This line may be due to dimethyl sulfide which is known to give an NMR line at δ = 2.058 in carbon tetrachloride (Ref. 53). The DMSO methyl-proton line appears at δ = 2.511 (Ref. 53). Both lines due to impurities are removed by treatment with calcium hydride.

(D) Acidic and basic impurities.
Acidic and basic impurities in DMSO have been determined by potentiometric titration of the solvent with dimsyl ion and with p-toluenesulfonic acid solutions in DMSO (Ref. 6 and 45).

(E) Conductivity.
The lowest specific conductance reported for DMSO is 2×10^{-8} ohm^{-1} cm^{-1} (Ref. 5). Since water ionizes more readily than DMSO, the specific conductance will be a measure of the concentration of both water and other ionizable substances.

TOXICITY AND OTHER HAZARDS

Buckley (Ref. 54) points out that DMSO dehydrates and defats the skin, but that apart from this hazard the solvent itself seems to be relatively free from toxic effects. It is the ability of DMSO to penetrate the skin rapidly and be incorporated into the bloodstream (as evidenced by the fact that soon after skin contact it can be tasted) that poses a potential hazard. Studies indicate (Ref. 54) that, in many cases, solutes are also carried along with the solvent through the skin. Thus, it is possible that a toxic substance could be more hazardous when in solution in DMSO. It is, therefore, advisable to take all necessary safety precautions when working with DMSO solutions in general and especially those of toxic substances.

Several substances have been reported to have caused explosions when mixed with DMSO (Ref. 24). These include sodium hydride (Ref. 55), periodic acid (Ref. 56), magnesium perchlorate (Ref. 57) and 70% perchloric acid (Ref. 57 and 24). It was also found (Ref. 58) that addition of DMSO to iron(III) perchlorate (G. F. Smith, non-yellow) is highly exothermic with explosion potential.

REFERENCES

1. H. H. Szmant, in S. W. Jacob, E. E. Rosenbaum, and C. D. Wood, eds., "Dimethyl Sulfoxide," Marcel Dekker, New York, 1971, Ch. 1.
2. D. Hadzi, J. Chem. Soc., 5128 (1962).
3. I. M. Kolthoff, M. K. Chantooni, Jr., and S. Bhowmik, J. Amer. Chem. Soc. 90, 23 (1968).
4. C. D. Ritchie, in J. F. Coetzee and C. D. Ritchie, eds., "Solute-Solvent Interactions," Vol. 1, Marcel Dekker, New York, 1969, Ch. 4.
5. I. M. Kolthoff and T. B. Reddy, Inorg. Chem. 1, 189 (1962).
6. C. D. Ritchie and R. E. Uschold, J. Amer. Chem. Soc. 89, 1721 (1967).
7. C. D. Ritchie and R. E. Uschold, J. Amer. Chem. Soc. 89, 2752 (1967).
8. E. C. Steiner and J. D. Starkey, J. Amer. Chem. Soc. 89, 2751 (1967).
9. F. G. Bordwell and W. S. Matthews, J. Amer. Chem. Soc. 96, 1214, 1216 (1974).
10. M. K. Chantooni, Jr. and I. M. Kolthoff, J. Phys. Chem. 79, 1176 (1975).
11. I. M. Kolthoff and M. K. Chantooni, Jr., J. Amer. Chem. Soc. 98, 5063, 7465 (1976).
12. E. M. Arnett and D. R. McKelvey, J. Amer. Chem. Soc. 88, 2598 (1966).
13. V. Gutmann and G. Schober, Z. Anal. Chem. 171, 339 (1959).
14. I. M. Kolthoff and T. B. Reddy, J. Electrochem. Soc. 108, 980 (1961).
15. J. N. Butler, J. Electroanal. Chem. 14, 89 (1967).
16. J. N. Butler, in P. Delahay, ed., "Advances in Electrochemistry," Vol. 7, Interscience, New York, 1970, p. 77.
17. E. Buncel and H. Wilson, Adv. Phys. Org. Chem. 14, 133 (1977).
18. J. F. Coetzee, D. Frollini, C. G. Karakatsanis, E. J. Subak, Jr., and K. Umemoto, in G. Mamantov, ed., "Characterization of Solutes in Non-Aqueous Solvents," Plenum Press, New York, 1978, Ch. 1.
19. E. M. Arnett, T. C. Moriarity, L. E. Small, J. P. Rudolph, and R. P. Quirk, J. Amer. Chem. Soc. 95, 1492 (1973).
20. International Union of Pure and Applied Chemistry Information Bulletin 14B.
21. H. L. Schlafer and W. Schaffernicht, Angew. Chem. 72, 618 (1960).
22. T. B. Douglas, J. Amer. Chem. Soc. 70, 2001 (1948).
23. J. Kenttamaa, J. J. Lindberg, and A. Niessema, Suomen Kemistilehti 33B, 189 (1960).
24. J. A. Riddick and W. B. Bunger, "Organic Solvents," Wiley-Interscience, New York, 1970, pp. 466-467, 857-860.
25. R. A. Hovermale and P. G. Sears, J. Phys. Chem. 60, 1579 (1956).
26. J. Kenttamaa and J. J. Lindberg, Suomen Kemistilehti 33B, 32 (1960).
27. J. J. Lindberg, T. Kenttamaa and A. Nissema, Suomen Kemistilehti 34B, 156 (1961).
28. F. A. Cotton and R. Francis, J. Amer. Chem. Soc. 82, 2986 (1960).
29. H. P. Bennetto and E. F. Caldin, J. Chem. Soc. A, 2191 (1971).
30. W. S. MacGregor, "The Reaction Solvent Characteristics of Dimethyl Sulfoxide," Crown Zellerbach Corporation, Chemical Products Division, Camas, Washington, March 1967.
31. J. F. Casteel and P. G. Sears, J. Chem. Eng. Data 19, 196 (1974).
32. E. M. Kosower, "An Introduction to Physical Organic Chemistry," Wiley, New York, 1968, pp. 296-305.

33. V. Gutmann and R. Schmid, Coord. Chem. Rev. 12, 263 (1974).
34. R. L. Amey, J. Phys. Chem. 72, 3358 (1968).
35. F. A. Bovey, J. W. Longworth and F. P. Hood, unpublished results.
36. J. N. Butler, et al., "Purification and Analysis of Organic Nonaqueous Solvents," Air Force Cambridge Research Laboratories, AFCRL-70-0605, 1970, AD718109, p. 71.
37. P. G. Sears, G. R. Lester and L. R. Dawson, J. Phys. Chem. 60, 1433 (1956).
38. T. B. Reddy, Ph. D. Thesis, University of Minnesota, Minneapolis (1960).
39. J. Menashi, W. L. Reynolds, and G. Van Auken, Inorg. Chem. 4, 299 (1965).
40. G. G. Price and M. C. Whiting, Chem. and Ind. (Lon.), 775 (1963).
41. E. J. Corey and M. Chaykovsky, J. Amer. Chem. Soc. 84, 866 (1962).
42. E. J. Corey and M. Chaykovsky, J. Amer. Chem. Soc. 87, 1345 (1965).
43. D. L. Maricle and W. G. Hodgson, Anal. Chem. 37, 1562 (1965).
44. J. L. Roberts and D. T. Sawyer, J. Electroanal. Chem. 9, 1 (1965).
45. C. D. Ritchie, G. A. Skinner, and V. G. Badding, J. Amer. Chem. Soc. 89, 2063 (1967).
46. J. F. Coetzee and C. D. Ritchie, in J. F. Coetzee and C. D. Ritchie, eds., "Solute-Solvent Interactions," Vol. 1, Marcel Dekker, New York, 1969, Appendix, p. 607.
47. C. K. Mann, in A. J. Bard, ed., "Electroanalytical Chemistry," Vol. III, Marcel Dekker, New York, 1969, Ch. 2.
48. V. A. Molochko and G. M. Kurdyumov, Khim. Prom. (Moscow) 46, 156 (1970).
49. D. R. Cogley and J. N. Butler, J. Phys. Chem. 72, 1017 (1968).
50. D. R. Cogley and J. N. Butler, "Study of the Kinetics of Alkali Metal Deposition and Dissolution in Non-Aqueous Solvents," Air Force Cambridge Research Laboratories, AFCRL-68-0560, 1968, AD 681453.
51. A. S. Meyer and C. M. Boyd, Anal. Chem. 31, 215 (1959).
52. R. J. Jasinski and S. Kirkland, Anal. Chem. 39, 1663 (1967).
53. F. A. Bovey and G. V. D. Jiers, in L. Meites, ed., "Handbook of Analytical Chemistry," McGraw-Hill, New York, 1963, Section 7.
54. A. Buckley, J. Chem. Ed. 42, 674 (1965).
55. Chemical and Engineering News, 44, April 11, 48 (1966).
56. J. J. M. Rowe, K. B. Gibney, M. T. Yang, and G. G. S. Dutton, J. Amer. Chem. Soc. 90, 1924 (1968).
57. Chemical and Engineering News, 43, September 13, 62 (1965).
58. J. F. Coetzee and C. G. Karakatsanis, unpublished results, 1977.

Dimethylformamide

Prepared for publication by

Jean Juillard

N,N'-Dimethylformamide (DMF) is a good solvent for organic and to a lesser extent inorganic compounds. It is, together with dimethyl sulfoxide and acetonitrile, one of the most widely used of the so-called dipolar aprotic solvents. Owing to its fairly high relative permittivity (dielectric constant), it is a moderately dissociating solvent for electrolytes. Acid-base reactions as well as thermodynamic properties of electrolyte solutions have been studied by many authors. Contrary to the N-methylamides (see Chapters 11 and 12), it is a typically weakly associated solvent, as seen (Ref. 1) from dielectric studies (the Kirkwood g factor is about one at all temperatures).

Owing to its electron-donor character, DMF reacts with many acids. For example, Gutmann's donicity number (Ref. 2) is 27. Its polarographic range is quite large, e.g., 3.5 V at the dropping mercury electrode with 0.1 M Bu_4NClO_4 as supporting electrolyte (Ref. 3). It is therefore widely used as a solvent for electrochemical reactions, especially reductions.

Pure DMF is colorless and, at room temperature, odorless. It is subject to thermal as well as photochemical degradation. In presence of water, DMF is slowly hydrolized according to the equation:

$$HCON(CH_3)_2 + H_2O \rightarrow HCOOH + (CH_3)_2NH$$

Formic acid and dimethylamine are thus predominant impurities in DMF and determine the odor of the impure solvent. They are weakly acidic and weakly basic, respectively; therefore, partial ionization does occur:

$$HCOOH + (CH_3)_2NH \rightleftarrows HCOO^- + (CH_3)_2NH_2^+$$

which results in a buffered solution (pH 11) with an increase in the conductivity of the solvent.

Thermal degradation produces dimethylamine and carbon monoxide. Hydrogen (Ref. 4) and hydrogen cyanide (Ref. 5) have been identified among the products of the photochemical degradation of the solvent.

Strongly basic media are difficult to obtain in DMF; there is, to our knowledge, no substance behaving as a strong base in DMF. If autoprotolysis of the medium actually occurs, the anion of the solvent must be very unstable (Ref. 6). It has been claimed (Ref. 7 and 8) that the autoprotolysis constant is smaller than 10^{-25} but no definite value has yet been proposed.

Attention must be paid to the fact that DMF has toxic effects, particularly on the liver and kidneys; the threshold value for air has been fixed (Ref. 9) at 30 mg/m^3.

PURIFICATION OF DIMETHYLFORMAMIDE

Good quality DMF is commercially available. As noted by Vaughn (Ref. 10), spectrograde solvent is not always suitable for all purposes. As a consequence of hydrolysis, the residual water content of commercial DMF is frequently low (0.1%). Many procedures have been proposed and used for the purification of the solvent. Four types of successive operation can be distinguished: treatment with a drying agent, neutralization of basic or acidic impurities, careful distillation, and elimination of gaseous impurities.

1. Preparing water-free solvent. Although the boiling point of water is far from that of DMF it is not possible to obtain a dry solvent by distillation only.

One of the first methods proposed for preliminary drying (Ref. 11) was azeotropic distillation with about 10% by volume of dry benzene; the benzene-water azeotrope is removed by distillation at atmospheric pressure. To prevent decomposition the temperature is maintained below 80°C. Alternatively, molecular sieves can be used. The solvent is kept in contact for periods ranging from 1 to 4 days with 4 Å (Ref. 12-15) or 5 Å (Ref. 16) sieves which are removed and replaced from time to time. Ritchie (Ref. 17) recommmends the use of Linde AW-500 molecular sieves in 1/16-inch pellets. Studying drying efficiency, he finds that the water content is less than 18 ppm after 27 hours. Molecular sieves can be dried before use by heating in a quartz tube under a stream of argon at 375°C for 24 h (Ref. 18). Finally, a procedure which uses chromatographic purification through alumina has been described by Moe (Ref. 13) in some detail. "A column approximately 100 cm long and 5 cm wide will contain 1 kg of alumina, sufficient for the purification of about 10 l of DMF". After bubbling of pure nitrogen for several hours the DMF thus obtained is thought to be convenient for polarographic use.

In our opinion, these three types of operation can be considered only as a first step in drying the solvent, and mild chemical drying agents must also be used. These range from anhydrous BaO (Ref. 11 and 19) to $MgSO_4$ (Ref. 20), Na_2CO_3 (Ref. 6 and 20), or $CuSO_4$ (Ref. 21). Surprisingly good samples of DMF can be obtained using storage of solvent over these chemicals for at least 24 h. It has been recommended that the drying agent be changed at least twice and the container shaken, if not continuously, at least from time to time. It also has been recommended that such an operation be performed in a cold, dark room. As far as Na_2SO_4 or Na_2CO_3 is concerned, the resulting solvents are of about the same quality (Ref. 20). Little or no degradation of the solvent (as estimated through the concentration of dimethylamine) results from such treatment (Ref. 11).

Some of the more common drying agents react with the solvent itself to produce significant amounts of acidic or basic impurities. BaO, cited previously, belongs to this category if it is used at temperatures above 30°C (Ref. 11). Other reagents are potassium hydroxide, calcium hydride (Ref. 5 and 22) and phosphorus pentoxide (Ref. 17, 23 and 24). P_2O_5 is the most frequently used, CaH_2 is probably the most efficient. Prue and Sherrington (Ref. 23) have shaken DMF for three days with P_2O_5, adding each morning about 10 g of fresh reagent. Recently, drying of amides using Vitrid, sodium bis(methoxy-2-ethoxy)aluminohydride, has been recommended (Ref. 25). In DMF it allows attainment of very basic media (pH 30). However, distillation of the solvent from the mixture obtained has not been attempted and is probably very hazardous. Whatever the method used, it is important to proceed with these operations in a dark room or apparatus to prevent any photochemical degradation.

2. Neutralization. Depending on the drying agent used, it has been recommended that the basic or acidic impurities produced be neutralized, either by shaking with picric acid (Ref. 20) or KOH pellets (Ref. 24). This last treatment is particularly recommended after drying over P_2O_5 which generates formic acid. Such neutralization can be done either before or after a first distillation.

3. Distillation. The drying process can be further carried out during this operation. The DMF is refluxed and distilled form P_2O_5 or CaH_2. However, owing to a degradation process increased by heating, it is preferable first to decant the solvent and transfer it under dry nitrogen, and then to distill it at reduced pressure.

The quality of the final product is greatly affected by the care with which the distillation is carried out. It seems to be important to work under vacuum, with a darkened column, or in a pure nitrogen or argon atmosphere. As a rule, the temperature must be kept under 60°C; heating must be gentle and overheating avoided. Distillation in daylight results in the production of hydrogen cyanide (Ref. 5), particularly in the presence of CaH_2. No traces of HCN are detected if the operations are conducted in the dark.

Many types of distillation apparatus currently described in the literature do not seem to be very efficient. It is not surprising to note that the best quality DMF, if conductivity is accepted as a test of purity, has been obtained by Brummer (Ref. 12), who used only molecular sieves as drying agent, but carried out the distillation in a slow current of dry nitrogen at low pressure (2 torr) and an efficient column (1 meter packed with Fenske helices). The use of a long column (60 cm at least) with good packing and reflux is recommended. For example, Tanaka (Ref. 21) distilled DMF which had been dried over anhydrous $CuSO_4$ at a pressure of 5 torr through an adiabatic fractional distillation column which was 1.3 cm in diameter, 120 cm in length and packed with Helipack coils. Dry nitrogen was passed through the apparatus during the distillation; 60% of the distillate was collected. The conductivity was lower than 1×10^{-7} ohm^{-1} cm^{-1} (25°C). Boiling temperatures at various pressures are given in Table 1.

TABLE 1. Recommended values for physical constants of dimethylformamide at 25°C and 1 atmosphere (except where otherwise noted).

Boiling temperature	T_B	152.3°C (Ref. 47)
		79°C at 61-62 torr (Ref. 37)
		55-56°C at 25-26 torr (Ref. 40).
		34°C at 2-3 torr (Ref. 15)
Melting temperature	T_M	-61°C
Refractive index (Ref. 44)	n_D^{25}	1.42689
Relative permittivity	ε	37.0
Surface tension (Ref. 45)	σ	37.1 dyne/cm
Viscosity (Ref. 23)	η	0.00796 poise
Density	ρ	0.9440 g cm^{-3}
Molal volume	V	77.39 cm^3
Heat capacity at constant pressure (Ref. 44)	C_p	37.4 cal/mol
Cubic expansion coefficient	α_P	1.00×10^{-3} K^{-1} *
Adiabatic compressibility coefficient (Ref. 44)	β_S	6.1×10^{-5} atm^{-1}
Isothermal compressibility coefficient	β_T	6.3×10^{-5} atm^{-1}*

*Calculated from data in Ref. (12)

4. Elimination of gaseous impurities. A flow of pure dry nitrogen or argon is passed through the solvent for several hours in order to eliminate oxygen, carbon monoxide and carbon dioxide. Such a solvent can then be used for polarographic purposes. A more complete deaeration can be achieved using a vacuum line.

5. Conclusions and recommendations. As various authors used different starting materials, it is difficult to compare the efficiency of the various methods of purification. Comparison between different ways of treating the same batch of solvent can be found to our knowledge in only two papers (Ref. 11 and 20). Thomas and Rochow (Ref. 11) always used first azeotropic distillation with benzene and compared subsequent treatment with $MgSO_4$, BaO, alumina and triphenylchlorosilane, followed in each case by distillation. Comparisons were made in terms of specific conductance and water content. Barium oxide as well as alumina treatment meet rather well these two criteria and do not have any side effects, such as producing dimethylamine or HCN. Juillard (Ref. 20) compared drying with Na_2SO_3, Na_2SO_4 or molecular sieves with azeotropic distillation with benzene and distillation over P_2O_5. As far as conductivity and water content are concerned, the different batches of solvent thus obtained were of about the same quality, except that P_2O_5 has the disadvantage of promoting degradation of the solvent and thus of decreasing the efficiency of the distillation; therefore the use of P_2O_5 is not recommended. As a confirmation it can be noted that authors using P_2O_5 or CaH_2 as drying agents did not obtain purer solvents than those who employed BaO or Na_2CO_3 or even only molecular sieves.

It is therefore recommended that use be made first either of azeotropic distillation with benzene, as suggested by Thomas (Ref. 11), or of treatment with molecular sieves, as suggested by Ritchie (Ref. 17), and that the resulting DMF be then shaken with Na_2CO_3 or, better, with BaO for 1 or 2 days. After decantation the DMF is distilled twice under nitrogen (pressure <15 torr) using a 1-m column. All these operations must be carried out in the dark. After deaeration the solvent should be stored under nitrogen and used as soon as possible.

TESTS FOR PURITY

Owing to its various modes of degradation (hydrolysis, thermal and photochemical decomposition) the principal impurities found in DMF are dimethylamine, formic acid, hydrogen cyanide, carbon dioxide and carbon monoxide. To this list must be added water, oxygen, which is quite soluble, and impurities resulting from the purification process.

Conductivity. As stressed earlier, hydrolysis as well as decomposition results in ionic impurities: dimethylammonium formate, carbonate or cyanide. Thus, the conductivity of the solvent is a very good test of its purity.

Experimental conductivities recorded for DMF are always higher than those reported for other aprotic solvents such as ketones or nitriles. According to a rough estimate, the theoretical conductivity of the pure solvent should be below $10^{-13}\ \Omega^{-1}\ cm^{-1}$. In fact, conductivities obtained by the most careful workers are scarcely ever less than 10^{-7}. The best values have been reported to our knowledge, by Brummer (Ref. 12) who used for conductometric studies a solvent having conductivities varying from 2×10^{-8} to $5 \times 10^{-8}\ \Omega^{-1}\ cm^{-1}$. Values below 5×10^{-7} have been reported by numerous authors and any batch of DMF which is more conducting can be considered to be substantially impure.

Density. See below.

Water. Water can be titrated by the Karl Fischer (K-F) reagent. Kanatharan (Ref. 22) recommends that the titration be conducted slowly, since K-F reagent reacts only slowly with small amounts of water.

Usual procedures do not allow the determination of less than 10 ppm of water. According to Muroi (Ref. 26) it is possible to titrate as little as 0.2 ppm by increasing the sharpness of the end point via the following procedure: "Add a 10-30 ml sample to 25 ml MeOH containing 8% of a pyridine-SO_2 solution (320 g SO_2/l pyridine) and titrate potentiometrically with K-F reagent having a titre of 0.1-0.5 mg H_2O/ml". The use of DMF as a solvent for K-F reagent also has been advocated (Ref. 27).

Prue (Ref. 23) has titrated water in DMF using triphenylsilyl chloride, from which, according to Thomas (Ref. 11), hydrogen chloride is liberated quantitatively by water (amines or acids are thought to interfere); the HCl content is then estimated from the conductivity of the solution.

It is quite easy to prepare a solvent which contains less than 50 ppm of water. Very low concentrations (< 5 ppm) are more difficult to attain. The best value, less than 3 ppm, has been reported by Libbey and Stock (Ref. 28).

Dimethylamine. Colorimetric methods have been used by some authors. In our opinion, as long as the autoprotolysis constant of the solvent is not known, it is often not possible to say exactly what is basic and what is acidic in DMF. Kolthoff (Ref. 24) has used p-nitrophenol in the colorimetric determination of total basicity, but specific determinations would be preferable.

Thomas and Rochow (Ref. 11) have based the determination of the amine content on the fact that dimethylamine forms with 1-fluoro-2,4-dinitrobenzene a complex which absorbs in the visible spectrum at 3812 Å. Solvent prepared by Chang and Criss (Ref. 29) was found to contain less than 1 ppm of dimethylamine by this method.

Another spectrophotometric method which allows the determination of the dimethylamine content down to 2 ppm with an error of $\pm$ 10% has been proposed by Pribyl (Ref. 30); dimethyldithiocarbamate, which absorbs at 445 nm, is formed by adding carbon disulfide and copper(II) acetate in an ethanol-pyridine solvent.

Chromatography was thought by Butler (Ref. 18) not to be a reliable means of establishing the organic impurity content of the solvent since DMF can decompose or hydrolyze at high temperatures. Nevertheless, careful studies of the proper experimental conditions have been undertaken (Ref. 31 and 32). In the paper by Filippov (Ref. 32) it is shown that dimethylamine can be determined in DMF at levels as low as 1 ppm using tetrahydroxyethylenediamine as a stationary phase, Polysorb-1 as a solid support and a column temperature of 75°C.

Dimethylamine is not electroactive at mercury but can give coordination compounds with cations which will affect the course of electrochemical reductions.

Formic Acid. In contrast to dimethylamine, formic acid is electroactive. Kanatharan and Spritzer (Ref. 22) have attributed to formic acid two peaks, one cathodic, the other anodic, which appear in cyclic voltammograms of aqueous dimethylformamide. Alternating current polarography (Ref. 33) and, better, pulse polarography, can be used to estimate the formic acid content.

Formic acid can also be determined by titration with a base. Potentiometric titration is preferred since it allows determination of the dimethylammonium formate content as well. Megliskii (Ref. 34) has titrated potentiometrically formic acid, dimethylamine and dimethylammonium formate in DMF using two titrants, 0.1 M $HClO_4$ and 0.1 M KOH, both in alcohol. Such a method is suitable only for concentrations of the order of at least 100 ppm.

Hydrogen Cyanide. Trisler et al. (Ref. 5) reported the presence of HCN in DMF distilled over CaH_2 in natural light. Concentrations ranged from 10^{-5} M to 10^{-3} M. Spectrophotometric titrations can be carried out with 4-nitrobenzil, which reacts with cyanide ion to form a deep violet ion.

Oxygen. Oxygen is rather soluble in DMF. A study of oxygen solubility in relation to the oxygen content of the gas phase has been made by James (Ref. 35). When the gas phase was air and pure oxygen, the solubility was 2.2×10^{-3} and 3.1×10^{-3} M, respectively.

Oxygen is an electroactive impurity which interferes in polarography and other electrochemical processes. Two waves are observed (Ref. 36) with $E_{1/2} = -0.8$ and -2.8 V vs. the aqueous SCE; the first corresponds to the reduction of oxygen to superoxide:

$$O_2 + e^- \rightarrow O_2^{\cdot -}$$

and the second one to the reduction of superoxide to peroxide ion:

$$O_2^{\cdot -} + e^- \rightarrow O_2^{=}$$

James (Ref. 35) has proposed two methods for the determination of the oxygen concentration; polarography and the Winkler method. Polarographic measurements are made at -1.2 V vs. SCE, in order to ensure that the measured diffusion current is not influenced by a polarographic maximum. A modified Winkler method allows concentrations as low as 10 ppm to be determined. It depends upon quantitative oxidation of iodide ion to iodine, and is described in some detail (Ref. 35).

PHYSICAL PROPERTIES OF DIMETHYLFORMAMIDE

Numerical values of physical constants are highly dependent on the purity of the solvent. Consequently, important discrepancies are found in the literature. The present recommended values result from a careful examination of three aspects: probable accuracy of the measurements, consistency of the data of various authors at different temperatures, and purification of the solvent. Such a choice is subject to personal evaluation and it seems prudent to give also the other references.

Density. The density is probably also a good criterion of the purity of the solvent. Contamination with water increases the density (Ref. 23). The following values of the density at 25°C have been found (Ref. 23,8,37,12): 0.9439, 0.9440_2, 0.9441_5 and 0.9442 g cm^{-3}, respectively. Old values greater than 0.9443 frequently found in tables are probably too high. New work by Kawaizumi and Zana (Ref. 38) seems to indicate that the density of the pure solvent is lower. These authors obtain values ranging from 0.94360 to 0.94368. It is our feeling that these data are more accurate than previous ones but such a low value ($\rho = 0.94364 \pm 0.00004$) must be confirmed by others before being accepted.

Values at various temperatures other than those appearing in Table 2 have been given by Gopal and Rizvi (Ref. 39). At 20°C Saphon (Ref. 40) has obtained $\rho = 0.94878$ g cm^{-3}, in good agreement with the value in Table 2.

TABLE 2. Recommended values for physical constants of dimethylformamide at various temperatures.

		ρ, g cm^{-3}	η, poise	ε
Temperature	20°C	0.9488	0.00845	38
	30°C	0.9394	0.00746	36.1
	40°C	0.9298	0.00664	34.4
	50°C	0.9202	0.00598	32.8
Reference		12	49	1

Viscosity. Values other than those in Table 2 can be found in References (29) and (41). Prue's data at 25°C are confirmed by measurements reported by Ames and Sears (Ref. 42).

Relative permittivity. Data given by Bass and Cole (Ref. 1) are preferred to previous results (Ref. 43) of Leader and Gormley (36.71 at 25°C). The value reported at 25°C is interpolated from measurements at various temperatures. Data of Saphon (Ref. 40) are in good agreement with the value reported in Table 2 at 20°C (ε = 38.13).

Miscellaneous. Data at various temperatures concerning refractive index, surface tension and isothermal compressibility can be found in Ref. (44), (45) and (12), respectively. Other data concerning thermodynamic properties are reported in Ref. (39) and (44). Plots of vapor pressure, heat of vaporization, heat capacity, density, viscosity, surface tension and thermal conductivity for a large range of temperatures have been drawn up by Gallant (Ref. 46). The solubilities of some sixty substances in DMF have been tabulated (Ref. 50). Organic reactions in or with DMF have been summarized (Ref. 51).

REFERENCES

1. S. J. Bass, V. I. Nathan, R. M. Meigham and R. H. Cole, J. Phys. Chem. 68, 1509 (1964).
2. V. Gutmann, "Coordination Chemistry in Non-Aqueous Solutions," Springer Verlag, 1968, p. 152.
3. M. Bréant, M. Bazouin, C. Buisson, M. Dupin and J. M. Rebattu, Bull. Soc. Chim. France, 5065 (1968).
4. J. Kroh, E. Burzynska, Bull. Acad. Polon. Sci., Sc. Chim. 21, 289 (1973).
5. J. C. Trisler, B. F. Freasier and Shi-Ming Wu, Tetrahedron Letters, 687 (1974).
6. Ram Chand Paul, P. S. Guraya and B. R. Sreenathan, Indian J. Chem. 1, 335 (1963).
7. M. Bréant and G. Demange-Guérin, Bull. Soc. Chim. France, 2935 (1969).
8. E. Lolette and J. Juillard, J. Sol. Chem. 3, 127 (1974).
9. H. E. Stokinger, "Documentation of Threshold Limit Values," revised ed., American Conference of Governmental Industrial Hygienists, Cincinnati, Ohio, 1966.
10. J. W. Vaughn in "The Chemistry of Non-Aqueous Solvents," ed. J. J. Lagowski, Academic Press, New York, 1967, Vol. II, p. 243.
11. A. B. Thomas and E. G. Rochow, J. Amer. Chem. Soc. 79, 1843 (1957).
12. S. B. Brummer, J. Chem. Phys. 42, 1636 (1965).
13. N. S. Moe, Acta Chem. Scand. 21, 1389 (1967).
14. S. C. Chan and J. P. Valleau, Can. J. Chem. 46, 853 (1968).
15. D. A. Owensby, A. J. Parker and J. W. Diggle, J. Amer. Chem. Soc. 96, 2682 (1974).
16. J. N. Butler, J. Phys. Chem. 72, 3288 (1968).
17. C. D. Ritchie and G. H. Megerle, J. Amer. Chem. Soc. 89, 1447 (1967).
18. J. N. Butler, J. Amer. Chem. Soc. 92, 2602 (1970).
19. O. N. Bhatnager and C. M. Criss, J. Phys. Chem. 73, 174 (1969).
20. J. Juillard, J. Chim. Phys. 67, 691 (1970).
21. N. Tanaka, personal communication.
22. P. Kanatharan and M. S. Spritzer, Anal. Letters 6, 421 (1973).
23. J. E. Prue and P. J. Sherrington, Trans. Faraday Soc. 57, 1795 (1961).
24. I. M. Kolthoff, M. K. Chantooni, Jr., and H. Smagowski, Anal. Chem. 42, 1622 (1970).
25. M. Bréant, Journées d'Electrochimie, Rouen (France), 10 Avril 1975; M. Bréant and J. Georges, C. R. Acad. Sc. Paris, Série C 280, 33 (1975).
26. K. Muroi and M. Ono, Bunseki Kagaku 20, 975 (1971).
27. V. A. Klimova, F. B. Sherman and A. M. L'vov, Isvest. Acad. Nauk SSSR, Ser. Khim., 2599 (1967).
28. A. J. Libbey and J. T. Stock, Anal. Chem. 42, 526 (1970).
29. S. Chang and C. M. Criss, J. Sol. Chem. 2, 457 (1973).
30. N. Pribyl and J. Nedbalkova, Fresenius Z. Anal. Chem. 232, 261 (1967).
31. V. A. Zverev and G. A. Krestov, Isv. Vyssh. Uchebn. Zavedenii, Khimiya i Khim: Tekhnol, 963 (1968).
32. Yu S. Filippov and Ya A. Tsarfin, Zh. Anal. Khim. 26, 1644 (1971).
33. A. Francina, Thèse Docteur-ès-Sciences, Lyon (France), 1973, p. 72.
34. V. A. Meglitskii and N. n. Kvasha, Khim. Volokna, 70 (1971).
35. H. J. James and R. F. Broman, Anal. Chim. Acta 48, 411 (1969).
36. D. L. Maricle and W. G. Hodgson, Anal. Chem. 37, 1562 (1965).
37. C. M. French and K. H. Glover, Trans. Faraday Soc. 51, 1418 (1955).
38. Personal communication of numerical values used in the paper by F. Kawaizumi and R. Zana, J. Phys. Chem., 78, 1099 (1974).
39. Ram Gopal and S. A. Rizvi, J. Indian Chem. Soc. 43, 179 (1966).
40. S. Saphon and H. J. Bittrich, Z. Phys. Chem., Leipzig 252, 113 (1973).
41. Ram Gopal and P. P. Rastogi, Z. Phys. Chem. N. F. (Frankfurt) 69, 1 (1970).
42. D. P. Ames and P. G. Sears, J. Phys. Chem 59, 16 (1955).
43. G. R. Leader and J. F. Gormley, J. Amer. Chem. Soc. 73, 5731 (1951).
44. B. E. Geller, Zh. Fiz. Khim. 35, 2110 (1961).
45. R. A. Stairs, W. T. Rispin and R. C. Makhija, Can. J. Chem. 48, 2755 (1970).
46. R. W. Gallant, Hydrocarbon Processing 48, 199 (1969).
47. B. V. Ioffe, Zh. Obshch. Khim. 25, 902 (1955).
48. J. R. Ruhoff and E. E. Reid, J. Amer. Chem. Soc. 59, 4012 (1937).
49. L. R. Dawson and W. W. Wharton, J. Electrochem. Soc. 107, 700 (1960).
50. Ram Chand Paul and B. R. Sreenathan, Ind. J. Chem. 4, 382 (1966).
51. R. S. Kittila, "Dimethylformamide: Chemical Uses," E. I. DuPont de Nemours and Co., 1967.

Hexamethylphosphoramide

Prepared for publication by

T. Fujinaga, K. Izutsu and S. Sakura

Hexamethylphosphoramide [hexamethylphosphoric triamide, tris(dimethylamino)phosphine oxide, hexametapol, usually abbreviated MHPA or HMPT] (see formula I) is a colorless, mobile, dipolar aprotic solvent with a somewhat unpleasant odor. It is a liquid over a wide range of temperatures and has a relatively high relative permittivity (dielectric constant). It owes its importance mainly to the fact that it is one of the most basic of the aprotic solvents.

PHYSICAL AND CHEMICAL PROPERTIES OF HMPA

Some physical properties of MHPA are summarized in Table 1; further details are in Ref. 2 and 8.

TABLE 1. Physical properties of hexamethylphosphoramide.

Property	Value	Ref.
Freezing temperature, °C	7.20	1
Boiling temperature, °C		
-at 760 torr	235	1
-at 15 torr	115	1
-at 1 torr	68-70	1
Refractive index, n_D		
-at 20°C	1.4584	2
-at 25°C	1.4564	2
Density, g cm^{-3} at 25°C	1.020	2-4
Viscosity, cP at 25°C	3.25	3,4
	3.23	5
Dipole moment, D at 25°C		
-liquid HMPA	5.38	2
	4.31	6
-HMPA in benzene	5.54	2
Relative permittivity		
-at 20°C	30.5	2
-at 25°C	29.6	2,3
	30.0	4
Electrolytic conductivity, ohm^{-1} cm^{-1}		
x 10^8 (lowest values obtained)	0.5 - 2	3
	2-4	7

The HMPA molecule has a pyramidal structure. Its P-O bond is polar with approximately 50 per cent ionic character and can be described as IIA or IIb (Ref. 1). The symmetrical distribution

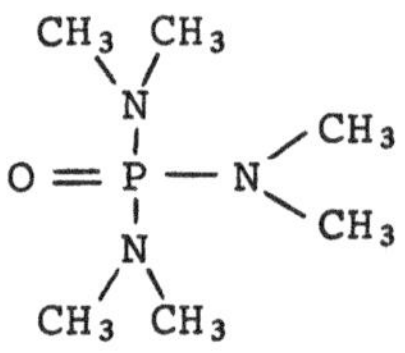

I

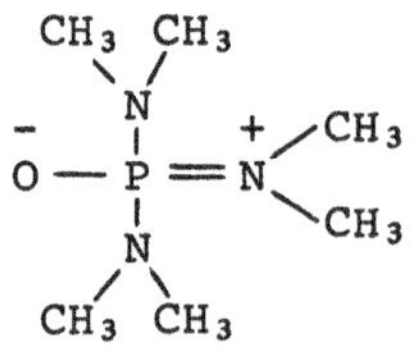

IIa

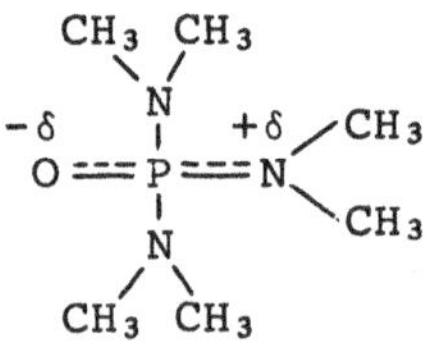

IIb

of the positive charge over the $\mathrm{N{-}P}\genfrac{}{}{0pt}{}{\diagup N}{\diagdown N}$ grouping and the high electron density on the oxygen give rise to a large dipole moment and unusually high basicity (Ref. 1). The strong basic character of HMPA, as compared with other solvents, is shown in Table 2 by the chemical shift for the proton of chloroform and by the donor number proposed by Gutmann.

TABLE 2. Parameters comparing basic strengths of hexamethylphosphoramide and other solvents.

Solvent	Donor number[a]	Chemical Shift[b]
Hexamethylphosphoramide	38.8	2.03
Pyridine	33.1	
Dimethyl sulfoxide	29.8	1.34
Dimethylformamide	26.6	1.30
Water	18.0	
Acetone	17.0	0.94
Propylene carbonate	15.1	
Acetonitrile	14.1	

[a]Ref. 9; see Chapter 1.

[b]Difference in NMR chemical shifts of chloroform at infinite dilution in the solvent in question and in cyclohexane; from Ref. 1.

HMPA is very hygroscopic and gradually takes up moisture if exposed to a humid atmosphere. It is easily mixed with water in any proportions.

HMPA is miscible in all proportions with a majority of the common solvents. It readily dissolves many organic compounds, some inorganic salts, polymers, and gases such as acetylene (Ref. 10). HMPA has strong solvation power for cations, and many metal complexes with it have been isolated. Most of the ionic compounds dissolved in it are fairly extensively dissociated (Ref. 3 and 11) because of its relatively high relative permittivity.

In HMPA, free from dissolved oxygen and moisture, alkali and alkaline earth metals dissolve to produce deep blue solutions of solvated electrons (Ref. 12), which are relatively stable even at ambient temperatures. Organic radical anions are also stable in HMPA (Ref. 13).

HMPA is fairly stable to electrochemical reduction, but it is not stable against anodic oxidation. In a HMPA-0.1 M lithium perchlorate solution, the potential range extends from +0.8 to -3.6 V *versus* Ag/0.01 M Ag^+ at a platinum electrode (Ref. 14 and 15) and from +0.1 to -2.8 V at a mercury electrode (Ref. 7 and 15) when measured at current densities of *ca.* 4 mA/cm^2. At the platinum electrode, the cathodic potential limit in this solution is determined by the generation of solvated electrons from the electrode (Ref. 16). In HMPA-0.05 M Bu_4NClO_4, the cathodic potential limit at a mercury electrode is -3.3 V *versus* Ag/0.01 M Ag^+ (Ref. 7 and 15). Metal ions which are strongly solvated in HMPA are less easily reduced than in other solvents and their reduction behavior is significantly influenced by the cation of the supporting electrolyte (Ref. 7 and 17).

In HMPA, perchloric, nitric and p-toluenesulfonic acids act as strong acids. The values of pK_a of benzoic acid, acetic acid and phenol have been reported by Madic and Trémillon (Ref. 18) to be 8.25, 11.9* and 14.15, respectively, which are several units larger than those in aqueous solutions.

Under ordinary conditions of storage, HMPA is very unstable. For the electrochemical use of HMPA, the degradation of the purified solvent resulting in the rapid formation of its peroxide (III) is most troublesome. The formation of the peroxide has been studied in detail by Gal and Yvernault (Ref. 11, 19 and 20). Oxygen and light are responsible for its formation. When freshly purified HMPA is exposed to the air and the light in a room for a few hours, a reduction wave as shown in Figure 1 appears in the polarographic residual current/potential curves. The rate of the degradation, however, is quite low when HMPA is kept either free from dissolved oxygen or in the dark.

*This value was obtained from the halfwave potential for the reduction of acetic acid at a platinized platinum electrode and should be checked.

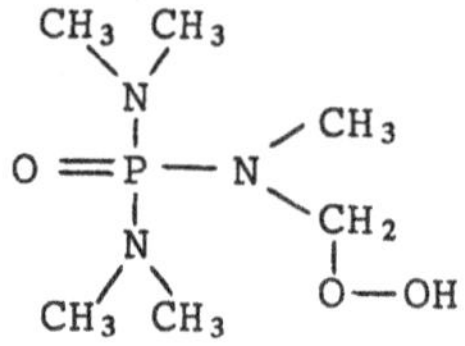

III

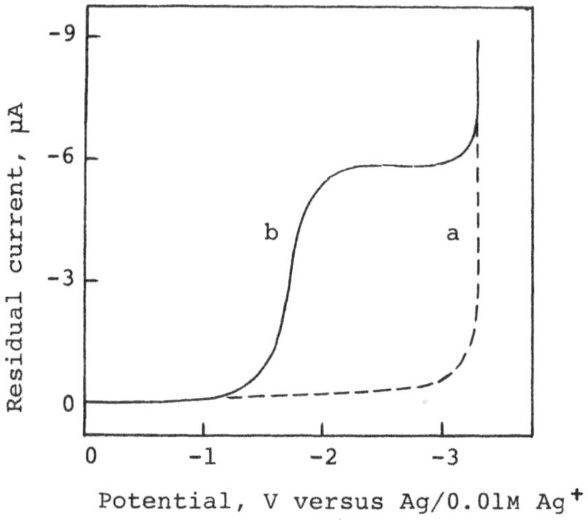

Figure 1. Polarograms of a 0.1 M Et_4NClO_4-HMPA solution at the dropping mercury electrode: (a) With freshly distilled HMPA and (b) with the same solvent kept for four days under light. From Ref. 11.

The decomposition of HMPA at higher temperatures is not particularly extensive, and appreciable decomposition products are not detected by gas chromatographic and infrared spectrophotometric measurements after the solvent has been heated for five hours in air at 100°C. If heated at 150°C, however, about one per cent of HMPA is decomposed (Ref. 21).

HMPA irritates the skin on contact for 24 hours or more, and the skin may become inflamed after prolonged use for more than a month. Kimbrough and Gaines (Ref. 22.a) found the amount of HMPA fatal to 50 per cent of test male and female rats to be 2,650 and 3,360 mg/kg, respectively, by oral administration. By dermal administration the corresponding amount was about 4000 mg/kg for both sexes. Testicular atrophy in male rats was noted at a single dose of 1000 mg/kg. Prolonged dietary intake of 40-80 mg/kg produced partial or complete atrophy. It has been reported that malignant tumors had been produced in laboratory animals by exposure to HMPA (Ref. 22b). Experiments with HMPA should therefore be carried out with care.

TESTS FOR PURITY

HMPA is usually synthesized industrially by the treatment of phosphorus oxychloride with an excess of dimethylamine in isopropyl ether. The reaction is

$$POCl_3 + 6HN(CH_3)_2 \rightarrow [(CH_3)_2N]_3PO + 3HCl.HN(CH_3)_2$$

The major impurities commonly found in HMPA as received are water, dimethylamine and its salt with hydrochloric acid, which come from the synthetic process. The peroxide of HMPA and other unidentified impurities are also found. The following tests can be employed to detect impurities in HMPA.

(a) Water content by the Karl Fischer method. The classical Karl Fischer titration, which needs a considerable amount of sample HMPA, gives unreliable results for smaller water contents, because HMPA forms a complex with iodine and a sharp end point cannot be obtained in the titration (Ref. 3). Gal and Yvernault (Ref. 15) obtained better results by using the coulometric method of Bizot (Ref. 23), in which only 0.05-0.2 ml of sample HMPA, negligibly small compared with the volume of the titration cell, was necessary.

(b) Determination of dimethylamine and chloride ion. The content of dimethylamine and chloride ion can be determined after the addition of water to the sample HMPA. Dimethylamine is determined by acid-base titration using a glass electrode. The chloride ion is titrated by conventional argentometry or preferably by coulometry at a silver anode. According to Gal and Yvernault (Ref. 15), the crude HMPA (product of Prolabo) generally contained at least 1-3 x 10^{-2} M chloride ion as determined coulometrically after being dissolved in water.

(c) Determination of the peroxide of HMPA by iodometry. Gal and Yvernault (Ref. 20) proposed an iodometric method in the presence of acetic acid. The mixture consisting of 1 g of potassium iodide, 5 ml of 0.5 M acetic acid and 1-2 ml of HMPA is warmed at 40°C for 15 min, and the liberated iodine is then titrated with a solution of 10^{-2} M sodium thiosulfate. It is preferable to add 20 ml of alcohol before titration and to determine the equivalence point by controlled-current potentiometry with two platinum indicator electrodes. Formaldehyde, which is also liberated, does not interfere with this titration.

(d) Polarographic measurements.* The measurement of polarographic residual current/potential curves is very useful as a test for electroactive impurities. The potential range applicable in pure HMPA has been described above (for details, see Ref. 15).

Dimethylamine and chloride ion give anodic waves at the dropping mercury electrode (DME). The decomposition product of the solvent, which was identified as the peroxide, can also be detected and determined by the reduction wave at $E_{1/2}$ = -1.75 V *versus* Ag/0.01 M Ag^+ at the DME in 0.1 M Et_4NClO_4 (Figure 1). Gal and Yvernault (Ref. 20) obtained a linear relation between the limiting current of the wave and the concentration of the peroxide. The presence of the peroxide also tends to shift the anodic potential limit in the positive direction, by forming a slightly conductive coverage of mercury oxides on the DME surface. The solvated proton and weakly acidic impurities can also be detected by their reduction waves. At the platinum electrode, the cathodic potential limit in sodium perchlorate-HMPA solutions is very sensitive to traces of water and shifts to less negative values in its presence (Ref. 14).

(e) Spectrophotometric measurements. Pure HMPA shows little or no absorption in the u.v. region between 260 and 400 nm, but impure HMPA very often absorbs in this region. Therefore, the recording of an u.v. spectrum in this region is a convenient test of the purity of HMPA, though the impurities that give the absorption have not yet been identified.

(f) Electrolytic conductivity measurements. Purified HMPA has an electrolytic conductivity of less than 4 x 10^{-8} ohm^{-1} cm^{-1}. Its electrolytic conductivity, however, increases fairly rapidly after distillation as a result of its decomposition (for the effect of the peroxide and dimethylamine on conductivity, see Ref. 26). This method is only useful for the detection of ionic impurities.

(g) Gas chromatographic method. A well-defined gas chromatographic peak of HMPA itself is obtained by using a column of the stationary phase of Silicone SE-30 at 150°C and a thermal conductivity or a flame ionization detector (Ref. 11 and 27). The decomposition products in relatively large amounts can be detected by this gas chromatogram, but this method is not useful for the test of trace impurities.

PURIFICATION AND STORAGE OF HMPA

Purification of HMPA is usually carried out by chemical pretreatment followed by distillation under reduced pressure or by recrystallization.

For ordinary electrochemical use, the following two procedures seem to be satisfactory and can be recommended.

(a) The purification procedure used by the present authors is as follows. The raw HMPA (product of the Nippon Oil Seal Ind. Co. Ltd, during 1970-74)is refluxed over barium oxide at about 4 torr in an atmosphere of nitrogen for several hours and then distilled. The middle fraction, about 70 per cent, is collected (bt *ca.* 90°C). The distillate is refluxed again over metallic sodium (*ca.* 1 g/kg) under reduced pressure (*ca.* 4 torr) and distilled. The distillation is carried out with an adiabatic vacuum fractional distillation column of 1 m length and packed with stainless steel Dixon packing. The distillate is kept in a nitrogen atmosphere or under vacuum in the dark at dry-ice temperature. The HMPA thus obtained gives an electrolytic conductivity of 2-4 x 10^{-8} ohm^{-1} cm^{-1} and a water content of approximately 0.01 per cent. This solvent is pure enough for both polarographic and u.v. spectrophotometric experiments.

*Perchlorates, very often used in polarography in nonaqueous solvents, are explosive at high temperatures, especially when heated in the presence of organic substances. Therefore, in the purification of perchlorates special care must be taken. Tetra-alkylammonium perchlorates should be dried at 50-70°C under vacuum (Ref. 24). Direct flames or infrared lamps should be avoided. It is advisable to purify only small amounts of the salt in one experiment (Ref. 25).

(b) Gal and Yvernault (Ref. 11 and 15) purified the raw HMPA (product of Prolabo) by two distillations under reduced pressure (<0.5 torr) in an atmosphere of argon, first over sodium (2 g/kg) and then over calcium hydride, using an adiabatic vacuum fractional distillation column of 50 cm length and packed with Multiknit. In order to avoid overheating, the distillation flask was heated in an oil bath. The yield was about 80 per cent. The purified solvent was kept under argon in the dark and below 7°C. According to Gal and Yvernault (Ref. 15), the treatment over sodium reduces the chloride content and that over calcium hydride reduces the dimethylamine. The purified HMPA contained 0.012-0.018 per cent of water, 0.5-4 x 10^{-3} M of dimethylamine and below 10^{-5} M of chloride ion, and the conductivity was about 7 x 10^{-8} ohm^{-1} cm^{-1}.

Dubois and others (Ref. 14) carried out the distillation under reduced pressure (1 torr) in the presence of sodium and triphenylmethane and then distilled again under an argon atmosphere. In order to obtain solvent for the voltammetric study of solvated electrons, Kanzaki and Aoyagui (Ref. 16) refluxed HMPA over calcium hydride for about ten hours in an atmosphere of nitrogen, and then distilled it under vacuum and collected the middle fraction (88°C/4 torr). The purified solvent was then introduced into a vacuum line (*ca.* 10^{-5} torr). After being degassed very carefully, the solvent was refluxed again over calcium hydride for about one month to remove moisture. Then it was distilled and transferred in a vacuum into ampoules with a break-seal. For electrolytic conductivity measurements, Hanna *et al.* (Ref. 3) obtained HMPA of conductivity of 0.5-2 x 10^{-8} ohm^{-1} cm^{-1}. Laboratory reagent grade HMPA (product of Pierrefitte Chimie) was distilled at a pressure of 0.8 torr through a heated fractionating column packed with small glass rings and the middle fraction of about 80 per cent was collected over calcium hydride. This fraction was then distilled again from calcium hydride, and a similar fraction was collected over barium oxide from which the solvent was distilled before use. All transfers were done under an atmosphere of dry argon.

By a recrystallization procedure, Luehrs and Leddy (Ref. 17) could reduce the polarographic background current at potentials more negative than -1 V. HMPA (Research Organic/Inorganic Chemical Corp.) was purified by fractional distillation under vacuum from calcium hydride at 60°C, and then recrystallized twice in a cold room at 0°C. In order to initiate crystallization, it was necessary to seed the HMPA with a crystal obtained by freezing HMPA with liquid nitrogen. About two thirds of the HMPA was allowed to freeze and the remaining liquid was decanted off. Other purification procedures are summarized in Ref. 9 and 28.

To reduce the water content in the purified HMPA, it is effective to pass the solvent through a column of molecular sieves. Gal *et al.* (Ref. 6) utilized a column of 3A molecular sieves and reduced the water content to about 0.005 per cent. To avoid formation of the peroxide, dry argon was passed over the column for half an hour before passing the solvent and the treatment was carried out in the dark. This treatment also achieved some reduction in the contents of dimethylamine and peroxide. However, extra care is needed when traces of conducting impurities leached out of the molecular sieves interfere with the experimental results.

The rate of decomposition of the purified HMPA may be effectively minimized by storing the HMPA in a cold dark place under an inert atmosphere or under a vacuum. It is, however, advisable to use the purified HMPA as soon as possible (in a day or two) after purification.

REFERENCES

1. (a) H. Normant, *Angew. Chem.* *79*, 1029 (1967); *Angew. Chem. Internat. Ed. Engl.* *6*, 1046 (1967).
   (b) H. Normant, *Bull. Soc. Chim. France*, 791 (1968).
2. (a) J. E. Dubois and G. Viellard, *J. Chim. Phys.* *62*, 699 (1965).
   (b) J. Y. Gal and C. Moliton-Bouchetout, *Bull. Soc. Chim. France*, 464 (1973).
3. E. M. Hanna, A. D. Pethybridge, J. E. Prue and D. J. Spiers, *J. Soln. Chem.* *3*, 563 (1974).
4. Y. M. Kessler, M. G. Formichev, N. M. Alpatova and V. P. Emelin, *Zh. Strukt. Khim.* *13*, 517 (1972).
5. T. Fujinaga, K. Izutsu and S. Sakura, *Nippon Kagaku Kaishi*, 191 (1973).
6. M. Schafer and C. Curran, *Inorg. Chem.* *4*, 623 (1965).
7. K. Izutsu, S. Sakura and T. Fujinaga, *Bull. Chem. Soc. Japan* *45*, 445 (1972); *46*, 493 (1973); K. Izutsu, S. Sakura, K. Kuroki and T. Fujinaga, *J. Electroanal. Chem.* *32*, App. 11 (1971).
8. J. A. Riddick and W. B. Bunger, "*Organic Solvents*," 2nd ed., Wiley-Interscience, New York, 1970.
9. V. Gutmann, "*Co-ordination Chemistry in Non-aqueous Solutions*," Springer, Vienna, 1968.
10. For example, T. Cuvigny and H. Normant, *Bull. Soc. Chim. France*, 2000 (1964).
11. J. Y. Gal, *Thesis*, University of Limoges (1972).
12. (a) T. Cuvigny, J. Normant and H. Normant, *CR Acad. Sci., Paris*, *C258*, 3502 (1964).
    (b) G. Fraenkel, S. H. Ellis and D. T. Dix, *J. Amer. Chem. Soc.* *87*, 1406 (1965).
    (c) N. Gremmo and J. E. B. Randles, *J. Chem. Soc., Faraday Trans. I*, 1480 and 1488 (1974).
13. M. Szwarc, *Accounts Chem. Res.* *2*, 87 (1969); *5*, 169 (1972).

14. J. E. Dubois, P. C. Lacaze and A. M. deFiquelmont, CR Acad. Sci., Paris, C262, 181, 249 (1966).
15. J. Y. Gal and T. Yvernault, Bull. Soc. Chim. France, 2770 (1971).
16. Y. Kanzaki and S. Aoyagui, J. Electroanal. Chem. 36, 297 (1972).
17. D. C. Luehrs and D. G. Leddy, J. Electroanal. Chem. 41, 113 (1973).
18. C. Madic and B. Trémillon, Bull. Soc. Chim. France, 1634 (1968).
19. J. Y. Gal and T. Yvernault, CR Acad. Sci., Paris, C275, 379 (1972).
20. J. Y. Gal and T. Yvernault, Bull. Soc. Chim. France, 839 (1972).
21. T. Takaoka, J. Synth. Org. Chem. Japan, 24, 412 (1966).
22. (a) R. Kimbrough and T. B. Gaines, Nature, London, 211, 146 (1966).
(b) N. V. Steere, J. Chem. Educ. 53, A12 (1976).
23. J. Bizot, Bull. Soc. Chim. France, 151 (1967).
24. C. K. Mann, in "Electroanalytical Chemistry," ed. A. J. Bard, Vol. III, Marcel Dekker, New York, 1969, p. 57.
25. W. C. Wolsey, J. Chem. Educ. 50, A 335 (1973).
26. J. Y. Gal, C. Moliton-Bouchetout and O. Perrier, Bull. Soc. Chim. France, 1212, 1972.
27. Nippon Oil Seal Ind. Co. Ltd., private communication.
28. H. Brusset, P. Delvalle, J. Garcin and P. Rajaonera, Bull. Soc. Chim. France, 3800, 1969.

Pyridine

Prepared for publication by

Meera Asthana and L. M. Mukherjee

INTRODUCTION

Only the most important properties of pyridine will be indicated here; further information is given in Ref. 1. Pyridine is a colorless liquid with freezing and boiling temperatures of -41.71°C (Ref. 2) and 115.3°C (Ref. 1 and 3) and a distinctive and disagreeable odor. It is a stable heterocyclic base with a pK_b in water of 8.81 at 25°C (Ref. 4). It resembles benzene in aromaticity and structural dimensions; due to the presence of the heteroatom, however, the planar molecule is distorted to some extent from the regular hexagonal shape of the benzene molecule (Ref. 5). The density and viscosity of the substance at 25°C are, respectively, 0.97824 g ml^{-1} and 0.884 cP (Ref. 6). Pyridine is considered to be truly aprotic in nature (Ref. 7), and it has only a negligible tendency for self-association in the liquid state (Trouton's constant = 21.7). The relative permittivity of the substance is 12.3 (Ref. 8) and its refractive index (n_D) is 1.50745 (Ref. 6) at 25°C. Burgess and Kraus (Ref. 9) estimated the specific conductance to be <1 x 10^{-9} ohm^{-1} cm^{-1} at 25°C. (Comprehensive lists of the physical constants are given in Ref. 1 and 10.)

Pyridine is miscible with the common organic solvents as well as water. It is known to form numerous binary, ternary, and quaternary azeotropes with water and organic liquids (Ref. 11 and 12).

From the time of its discovery, pyridine has been produced chiefly by high-temperature pyrolysis; feedstocks included bone oil, petroleum, and coal. Purification was tedious; from 15 tons of coal only 150 ml of pure pyridine was realizable (Ref. 13). Not until the late 1950's did synthetic pyridine become commercially available. Since that time a variety of efficient preparative procedures have been achieved. Catalytic dehydrogenation of piperidine over palladium at 260°C, for example, affords pyridine in 53% yield (Ref. 14); however, piperidine itself is not easily prepared, and the technique is thus of little commercial value.

More useful catalytic processes have been developed using various nitrogen and oxygen compounds to build the heterocycle. Most of the catalysts are based on silica and alumina and employ various additives. For example, an untreated silica-alumina catalyst (SAC) at 510°C converts 1-cyanobutadiene-1,3 to 35% crude pyridine (Ref. 15). At 300°C a gas stream containing ammonia, water and 3,4-dihydro-2-ethoxy-2H-pyran is converted to pyridine in 50% yield when passed over SAC treated with boric and phosphoric acids (Ref. 16); with a palladized SAC, 66% conversion is realized (Ref. 17). Crotonaldehyde, formaldehyde, water, air and ammonia are combined over untreated SAC, affording 67% pyridine and 7% of the picolines (Ref. 18). Ammonia is combined with water and tetrahydrofurfuryl alcohol at 500°C over a platinized SAC, producing pyridine in 48% yield (Ref. 19).

Other catalyst systems have been used. For example, the last reaction, when carried out over a molybdenum trioxide catalyst at 400°C, gives 69% of a pyridine fraction (Ref. 20). In another process, using a molybdenum-vanadium-alumina catalyst, methanol and 2,2'-dihydroxy-diethylamine are combined at 520°C, affording 42% pyridine and 10% picolines (Ref. 21).

METHODS OF PURIFICATION

It would seem that the likely impurities in commercial pyridine are water and amines such as the picolines and the lutidines. The literature abounds with purification procedures, most of which involve various preliminary chemical and physical treatments prior to distillation.

Pyridine and zinc chloride form an adduct which can be easily recrystallized from ethanol (Ref. 22); stable complexes of pyridine with mercuric chloride (Ref. 23) and with ethyl-oximinocyanoacetate (Ref. 24) are also known. The free base can be subsequently liberated by treating these addition compounds with concentrated sodium hydroxide solution (Ref. 22b). A drying step followed by distillation affords good product. Alumina, barium oxide, Drierite, sodium or potassium hydroxide, and calcium hydride have been used as drying agents.

Steam distillation of AR grade pyridine dissolved in 1.2 equivalents of 23% sulfuric acid has also been resorted to for removing contaminants (Ref. 2); the base was liberated and dried by using sodium hydroxide and was then fractionally distilled through a 50-plate Stedman column at a reflux ratio of 50:1. The product obtained was 99.5% pure (bt 93.7-94.0°C at 400 torr) with water as the major impurity; fractional freezing and redistillation of this material afforded 99.85 ± 0.03 mol % pyridine.

Repeated fractional freezing followed by refluxing over barium oxide and distillation gives reasonably pure pyridine, bt 114.5°C at 740 torr (Ref. 25).

High purity pyridine was obtained by distilling Baker & Adamson (ACS reagent grade) product in an oxygen-free atmosphere through an 11 ft x 1 in Helipak column. The lower-boiling pyridine-water azeotrope distilled first; later fractions (15% of the total) were 99.92% pure pyridine. These fractions were degassed, dried over calcium hydride and finally distilled under vacuum to provide standard samples of pyridine (Ref. 6).

Elving and co-workers have found AR grade (Baker, Mallinckrodt or Merck) pyridine which had been dried over Linde 4A molecular sieves for two to three days to be sufficiently pure for their polarographic work (Ref. 26 and 27). Through an adaptation of the fractional crystallization technique (cf. Ref. 25), Hall and Elving (Ref. 28) seem to have succeeded in obtaining a somewhat better product.

Recommended Procedure. The following method (Ref. 29) is recommended for obtaining pure pyridine.

(a) Keep Karl Fischer reagent grade or an equivalent grade of pyridine over solid potassium hydroxide (ca. 20 g/Kg) for two weeks.

(b) Fractionally distill the supernatant liquid from (a) over freshly activated Linde 5A molecular sieves and solid potassium hydroxide at a reflux ratio of 20:1, using a suitable column. Collect the fraction boiling at 115.3°C (760 torr).

The specific conductance of the above product should be between 3×10^{-8} and 5×10^{-8} ohm^{-1} cm^{-1} at 25°C. No water should be detectable by Karl Fischer titration. In 0.1 M $LiClO_4$ solution, the above product gave a pre-discharge wave (0.16 μA) at ca. -1.6 V, accompanied by two small waves (0.10 and 0.03 μA) at ca. -0.7 V and and -1.1 V, at a dropping mercury electrode between 0 V and the decomposition potential of -1.8 V vs. a mercury pool anode (Ref. 30).

Tests for Purity. Sulfur-containing impurities (Ref. 2) may be detected at less than 0.003 mol % by combustion analysis. Pyrrole, if present at levels above 1 ppm, may be detected by the formation of a noticeable coloration when an acidified sample of the pyridine is treated with p-dimethylaminobenzaldehyde (Ref. 2). Vapor phase chromatography with a column of 10% ethanolamine +5% o-phenylphenol on C-22 Firebrick at 90°C permits successful resolution of water, the picolines, 2,6-lutidine and pyridine (Ref. 31). Infrared spectroscopy has been used to determine water in pyridine down to 0.01 mol % (Ref. 32); Karl Fischer titration may also be used for water determination. Determination of total dissolved impurity in relatively pure samples of pyridine may be accomplished via the melting point curve technique (Ref. 33).

Storage of the Solvent and General Comments. To prevent contamination with atmospheric carbon dioxide and water, the solvent should be stored in a sealed container with an arrangement for dispensing it automatically, as required, under pressure of pure nitrogen gas.

Chronic exposure to pyridine vapor causes irritation of the mucous membranes and also affects the gastro-intestinal system; acute exposure may lead to unconsciousness (Ref. 34). (The odor of the material may be detected long before it becomes toxicologically important.) Pyridine should be used with adequate ventillation.

REFERENCES

1. R. Lindauer and L. M. Mukherjee, Pure Appl. Chem. 27, 267 (1971).
2. D. P. Biddescombe, E. A. Coulson, R. Handley, and E. F. G. Herington, J. Chem. Soc. 1957 (1954).
3. E. F. G. Herington and J. F. Martin, Trans. Faraday Soc. 49, 154 (1953).
4. N. F. Hall and M. R. Sprinkle, J. Amer. Chem. Soc. 54, 3469 (1932).
5. B. Bak, L. Hansen-Nygaard and J. Rastrup-Anderson, J. Molec. Spectros. 2, 361 (1958).
6. R. V. Helm, W. J. Lanum, G. L. Cook, and J. S. Ball, J. Phys. Chem. 62, 858 (1958).
7. I. M. Kolthoff, J. Polarogr. Soc. 10, 22 (1964).
8. A. A. Maryott and E. R. Smith, Circ. U.S. Nat. Bur. Stand. No. 514 (1951).
9. D. A. Burgess and C. A. Kraus, J. Amer. Chem. Soc. 70, 706 (1948).

10. L. M. Mukherjee, CRC Crit. Rev. Anal. Chem. 4, 325 (1975).
11. A. Galska-Krajewska and K. Zieborak, Roczn. Chem. 36, 119 (1962); Chem. Abstr. 57, 6680h (1962).
12. L. H. Horsely, "Azeotropic Data," Advances in Chemistry Series, Vol. 6, American Chemical Society, Washington, D.C., 1952, p. 135.
13. A. Albert, "Heterocyclic Chemistry," University of London, The Athlone Press, London, 1959, p. 101.
14. Farbwerke Hoechst A. G., German Pat. No. 1192648 (1965); Chem. Abstr. 63, 8323g (1965).
15. Japan Gas-Chemical Co. Ltd., Japanese Pat. No. 23175 (1965); Chem. Abstr. 64, 3502f (1966).
16. Imperial Chemical Industries Ltd., Brit. Pat. No. 1008455 (1965); Chem. Abstr., 64, 3496f (1966).
17. Imperial Chemical Industries Ltd., Netherlands Appl. 6408149 (1965); Chem. Abstr. 63, 2961b (1965).
18. Distillers Co., Ltd., Brit. Pat. No. 1038537 (1966); Chem. Abstr. 65, 16949f (1966).
19. Imperial Chemical Industries Ltd., Netherlands Appl. 6605492 (1966); Chem. Abstr., 66, 85703s (1967).
20. Distillers Co., Ltd., Netherlands Appl. 6409131 (1965); Chem. Abstr. 63, 584a (1965).
21. Ruetgerswerke und Teerverwertung A. G., Belgian Pat. No. 660964 (1965); Chem. Abstr. 64, 2065 f (1966).
22. (a) J. H. Mathews, J. Amer. Chem. Soc. 48, 562 (1926); (b) W. F. Luder and C. A. Kraus, J. Amer Chem. Soc. 69, 2481 (1947).
23. W. G. Schneider, H. J. Bernstein and J. A. Pople, Canad. J. Chem. 35, 1487 (1957).
24. B. Adcock, A. Lawson, and J. O. Stevens, Analyst 86, 615 (1961).
25. D. G. Leis and B. C. Curran, J. Amer. Chem. Soc. 67, 79 (1945).
26. A. Cisak and P. J. Elving, J. Electrochem. Soc. 110, 160 (1963).
27. (a) M. S. Spritzer, J. M. Costa, and P. J. Elving, Anal. Chem. 37, 211 (1965); (b) W. R. Turner and P. J. Elving, Anal. Chem. 37, 467 (1965); (c) J. E. Hickey, M. S. Spritzer, and P. J. Elving, Anal. Chim. Acta 35, 277 (1966).
28. D. A. Hall and P. J. Elving, Anal. Chim. Acta 39, 141 (1967).
29. L. M. Mukherjee, J. J. Kelly, W. Baranetsky and J. Sica, J. Phys. Chem. 72, 3410 (1968).
30. R. Lindauer and L. M. Mukherjee, unpublished work.
31. F. Kametani and S. Kabuta, Yakugaku Zasshi 82, 659 (1962); Chem. Abstr. 57, 6588h (1962).
32. E. A. Coulson, J. L. Hales, and E. F. G. Herington, J. Chem. Soc., 2125 (1951).
33. J. P. McCullough, D. R. Douslin, J. F. Messerlg, I. A. Hossenlopp, T. C. Kincheloe, and G. Waddington, J. Amer. Chem. Soc. 79, 4289 (1957).
34. E. Browning, "Toxic Solvents," Arnold, London, 1953, pp. 148-149.

Ethylenediamine

Prepared for publication by

Meera Asthana and L. M. Mukherjee

INTRODUCTION

Only the most important properties of ethylenediamine will be indicated here; further information is given in Ref. 1. Ethylenediamine (EDA) is a colorless, water-soluble basic liquid with an ammoniacal odor. Its freezing and boiling temperatures have been reported as 11.0°C (Ref. 2) and 11.3°C (Ref. 3), and 117.2°C (Ref. 4 and 5), respectively. Its pK_b in water at 25°C is 4.07 (Ref. 4). At 25°C, its density, viscosity and relative permittivity are 0.891 g cm^{-3} (Ref. 2), 1.54 cp (Ref. 6) and 12.9 (Ref. 7). At 20°C, its refractive index, n_D, is 1.4532 (Ref. 8) and the minimum value reported so far for its specific conductance is 9×10^{-9} ohm^{-1} cm^{-1} (Ref. 8). Its autoprotolysis constant is uncertain; values at 25°C of ca. 10^{-12} M^2 (Ref. 9) and 5×10^{-16} M^2 (Ref. 10) have been reported.

The substance is extremely hygroscopic; a monohydrate with a melting temperature of 10°C (Ref. 4) is known. Besides, it forms a solid carbamate, mt 170°C (Ref. 11), with carbon dioxide. Thus, the two common impurities in EDA are expected to be water and carbon dioxide, while the presence of other impurities in the commercial product would be chiefly determined by the method of preparation employed. EDA can be prepared by reacting ethylenedichloride with alcoholic (Ref. 12b) or aqueous (Ref. 13) ammonia. The latter is a high-pressure reaction and is reported to give low conversion of ammonia and large amounts of by-products. EDA has also been prepared from ethylenedibromide by phthalimide synthesis (Ref. 14). Certain organic chlorides such as ethylenedichloride, if present as contaminants, are known to react with the solvent (at 100°C) to produce secondary amines and the corresponding hydrochlorides (Ref. 15). Likely impurities formed as by-products during the manufacture of EDA include piperazine and similar cyclic materials, 2-ethylpyrazine and higher polyethylene amines, e.g., diethylenetriamine, triethylenetetramine and tetraethylenepentamine.

Companies such as Dow, Union Carbide, and Matheson, Coleman and Bell in the United States apparently produce their EDA through the high-pressure reaction between ethylenedichloride and ammonia mentioned above.

METHODS OF PURIFICATION

EDA cannot be dried by fractional distillation because it forms a constant-boiling (bt 118.5°C) fraction containing ca. 15% by weight of water. Details of the methods that have been used for the purification of EDA are given below.

Drying of EDA with sodium metal was first recommended by Hoffmann (Ref. 12a). Current practice involves the treatment of EDA with solid alkali (NaOH or KOH), then refluxing over molten sodium metal and, finally, fractional distillation in an atmosphere of nitrogen or hydrogen gas.

Several modifications of the above procedure exist. White and Morgan (Ref. 16) purified their EDA by three successive distillations from sodium. Putnam and Kobe (Ref. 2) treated their sample (70% EDA) with a large excess of sodium hydroxide on a steam bath; the EDA layer separating out at 60°C was decanted off and treated again with sodium hydroxide. Subsequently, the material was refluxed with sodium shavings and fractionated. Through an essentially similar procedure which consisted of heating the 70% sample with sodium hydroxide and then refluxing with 2-3% of sodium for 2-3 hours, Clarke and Bloat (Ref. 17) claimed to have obtained "100% EDA", as determined by titration with hydrochloric acid using methyl red as indicator. Bromley and Luder (Ref. 8) stored their starting material (95% EDA) over a mixture of sodium hydroxide and barium oxide for several days, kept it in contact with sodium metal at room temperature for a day and then fractionally distilled from a mixture of activated alumina and sodium hydroxide. The product so obtained was finally fractionated two more times from alumina. Schmidt and co-workers (Ref. 18) refluxed over lithium metal for several days and then used a double fractional distillation from sodium metal. Siefkar (Ref. 19) refluxed commercial EDA twice over sodium metal followed each time by a fractional distillation (at 250 torr, bt 85°C) in a current of nitrogen; the first 10% of the distillate was discarded in each distillation. This procedure seemed to have removed a polarographically active impurity ($E_{1/2}$ = 0.15 V vs. saturated $Zn(Hg)/ZnCl_2$ (s), LiCl (s) reference electrode) present in the starting material.

Schober and Gutmann (Ref. 8) have demonstrated that detectable quanties of impurities may be leached from glassware by EDA. The specific conductance of their EDA decreased each time as they repeated (five times) the distillation from sodium metal into vessels previously "steamed" with EDA vapor for several hours. Alternatively, receptacles were pre-soaked in EDA for two (or three) days. Their lowest specific conductance (9×10^{-9} ohm^{-1} cm^{-1}) was one-tenth of that reported by other workers (Ref. 7). Using lithium metal instead of sodium, Schober and Gutmann found that a larger number of distillations were required to arrive at the same value of specific conductance. On heating to 100°C with potassium metal, detectable amounts of ammonia were produced along with a dark-blue, highly viscous residue insoluble in water, ethanol and diethyl ether.

According to Bruckenstein and Mukherjee (Ref. 5), the commercially available EDA, both Dow (98%) and Eastman-Kodak (95-100%), contains impurities absorbing at 260, 265-270, and 310-312 nm in 95% ethanol. The commercial liquids as such did not transmit appreciably at wavelenths below 340 nm. Several methods of purification investigated by these workers are described below.

In the first method, EDA was fractionally distilled from sodium metal and the 114-117.5°C fraction was shaken overnight with freshly activated F20 ALCOA alumina (ca. 20 g/l); the alumina was allowed to settle and the supernatant liquid was then fractionally distilled in the presence of another batch of alumina. All distillations were carried out in a nitrogen atmosphere using a reflux ratio of 20:1. The fraction boiling at a constant temperature of 117.2°C at 760 torr indicated the presence of a minimum amount of the ultraviolet-absorbing impurity mentioned. In the second method, commercial EDA (Dow 98%) was shaken with Linde molecular sieves (Type 5A - 1/16 in. pellets - 70 g/l) and then with a mixture of 15 g potassium hydroxide and 50 g calcium oxide per liter of the solvent. This was followed by fractional distillation of the supernatant EDA from a batch of freshly activated molecular sieves in an atmosphere of nitrogen gas. The fraction boiling at 117.2°C at 760 torr was collected. The water content (as determined by Karl Fischer method) of the molecular sieve-treated sample was considerably lower (0.015 M vs. 0.05 M) than that obtained from the sodium treatment described above (Ref. 9). However, with the molecular sieve-treated EDA two polarographic waves were observed (Ref. 20) at potentials which correspond to +0.040 V (I) and -0.140 V (II) vs. the saturated $Zn(Hg)/ZnCl_2$ (s), LiCl(s) reference electrode of Schaap et al. (Ref. 10). A single fractional distillation of the molecular sieve-treated material from sodium metal completely removed the second impurity; however, distillation from calcium hydride almost doubled the wave height due to this particular impurity. Both sodium metal and calcium hydride did not appear to have any effect on the first wave. Moreover, changes in ultraviolet absorption at ca. 340 nm paralleled the polarographic observations in the present case. Also, no change in water content could be found on further distillation of the molecular sieve-treated sample from either sodium metal or calcium hydride.

Bruckenstein and Mukherjee (Ref. 9) have also fractionally distilled commercial EDA a number of times from sodium metal. Each successive distillation improved the ultraviolet spectrum.

Multiple extractions (Ref. 9) of the sodium-distilled EDA with fresh batches of purified cyclohexane showed that the ultraviolet-absorbing impurities were extractable. However, the following procedures did not produce any improvement in the spectrum of the sodium-treated solvent: (i) fractional crystallization, and (ii) treatment with sodium borohydride, sodium hydride, cation- and anion-exchange resins, silica gel, and subsequent distillation.

Azeotropic distillation has been used by Creamer and Chambers (Ref. 21) for removing water from EDA. They added benzene or isopropyl ether to form a low-boiling azeotrope with water. Benzene removed water faster than isopropyl ether. The method yielded products of at least 99.5% purity and having a specific conductance of no more than 1.0×10^{-5} ohm^{-1} cm^{-1}.

Recommended Procedure. A recommended procedure for obtaining pure EDA is the following:

(a) Shake the commercial sample (98% EDA) with activated molecular sieves (Linde Type 5A - 70 g/l) for approximately 12 hours.

(b) Decant the supernatant liquid from (a) and shake it for about 12 hours with a mixture of calcium oxide (50 g/l) and potassium hydroxide (15 g/l).

(c) Fractionally distill (20:1 reflux ratio) the supernatant liquid obtained from (b) in the presence of freshly activated molecular sieves. Collect the fraction boiling at 117.2°C at 760 torr ($\partial T/\partial P$ = 0.041°C/torr).

(d) The distillate obtained from (c) should be fractionated at least once from sodium metal.

All distillations should be carried out in an atmosphere of pure nitrogen gas. EDA thus purified has the following characteristics.

Boiling temperature (760 torr)	117.2 ± 0.1°C
Freezing temperature*	11.3°C
Heat of fusion*	64.5 ± 0.6 cal g^{-1} (11.3°)
Freezing temperature depression constant*	2.43 ± 0.02 K mol^{-1} Kg^{-1}
Ultraviolet absorption	Peak at 310-330 nm; absorbance (1-cm light path) = 0.25; cut-off: *ca.* 280 nm.
Polarographic impurity	$E_{1/2}$ = +0.04 V *vs.* sat. Zn(Hg)/$ZnCl_2$(s), LiCl(s) reference electrode (Ref. 10); i_d = 2 μA in saturated LiCl as a supporting electrolyte.
Water content	*ca.* 0.015 M

*EDA purified by molecular sieve treatment (steps a to c above) was used. Cryoscopic measurements (Ref. 3) showed that the total impurity content of the molecular sieve-treated EDA was 0.030 M; half of this was due to water (*ca.* 0.015 M). Most of the remaining impurity probably was dissolved nitrogen gas.

Storage of the Solvent and General Comments. To prevent contamination with the carbon dioxide and water from the air, EDA should be stored in a sealed reservoir with arrangement for dispensing it automatically, when required, under pressure of pure nitrogen gas.

Direct contact of EDA with the skin causes severe burns; also, EDA vapors are extremely irritating and may lead to bronchial complications. Adequate protection is absolutely necessary while handling the solvent.

REFERENCES

1. L. M. Mukherjee and S. Bruckenstein, *Pure Appl. Chem.* *13*, 421 (1966).
2. G. L. Putnam and K. A. Kobe, *Trans. Electrochem. Soc.* *74*, 609 (1938).
3. L. D. Pettit and S. Bruckenstein, *J. Inorg. Nucl. Chem.* *13*, 1478 (1962); F. A. K. Badawi, M.S. Thesis, University of Minnesota, 1964.
4. "*Lange's Handbook of Chemistry*," Handbook Publishers, Inc., Sandusky, Ohio, 1956, pp. 553, 1203.
5. S. Bruckenstein and L. M. Mukherjee, *J. Phys. Chem.* *64*, 1601 (1960); *ibid.* *66*, 2228 (1962).
6. A. E. Dunsten, T. P. Hilditch, and F. B. Thole, *J. Chem. Soc.* *103*, 133 (1913).
7. W. H. Bromley, Jr., and W. F. Luder, *J. Amer. Chem. Soc.* *66*, 107 (1944).
8. G. Schober and V. Gutmann, *Monatsh. Chem.* *89*, 649 (1958).
9. L. M. Mukherjee, Ph.D. Dissertation, University of Minnesota, 1961.
10. W. B. Schaap, R. E. Bayer, J. R. Siefkar, J. L. Kim, P. W. Brewster, and F. C. Schmidt, *Record Chem. Progr.* *22*, 197 (1961).
11. E. Schering, German Patent 123,138 (July 30, 1901).
12. A. W. Hoffmann, (a) *Proc. Roy. Soc.* (London) *10*, 224 (1859); (b) *Ber. dt Chem. Ges.* *4*, 666 (1871).
13. G. O. Curme and F. H. Lommen, U.S. Patent 1,832,534 (Nov. 17, 1931); Goodyear Tyre and Rubber Co., British Patent 393,093 (June 1, 1933).
14. S. Gabriel, *Ber. dt Chem. Ges.* *22*, 2224 (1889).
15. "*Union Carbide Corporation Laboratory Manual*" 31-9A-2 (March 16, 1956).
16. A. H. White and S. O. Morgan, *J. Chem. Phys.* *5*, 655 (1937).
17. J. T. Clarke and E. R. Bloat, *J. Polymer Sci.* *1*, 419 (1946).
18. B. B. Hibbard and F. C. Schmidt, *J. Amer. Chem. Soc.* *77*, 225 (1955).
19. J. R. Siefkar, Ph.D. Dissertation, Indiana University, 1960.
20. L. M. Mukherjee, unpublished data.
21. R. M. Creamer and D. H. Chambers, *J. Electrochem. Soc.* *101*, 162 (1954).

N-Methylacetamide

Prepared for publication by

L. A. Knecht

N-Methylacetamide (abbreviated NMA in this report) is one of a small group of nonaqueous solvents which have relative permittivities (dielectric constants) greater than that of water. (The value for NMA is 165.5 at 40°C: Ref. 1). This property, coupled with the relatively strong solvating power of NMA, results in high solubilities of inorganic salts as compared to those in most other organic solvents. For example, the alkali halides studied conductometrically by Dawson and his co-workers (Ref. 1) exhibit solubilities greater than 0.1 M. Likewise it is possible to prepare a 0.1 M solution of tetraethylammonium perchlorate or sodium perchlorate in NMA (Ref. 2).

PHYSICAL PROPERTIES OF NMA

The physical properties of NMA are summarized in Tables 1 and 2.

TABLE 1. Some physical properties of N-methylacetamide.

Property	Value	Ref.
Melting temperature, °C	29.5	1
	30.6	3
	30.35-30.55	4
Boiling temperature, °C at 1 atm.	206	1
Specific conductance, ohm^{-1} cm^{-1} at 40°C (lowest value)	$1\text{-}3 \times 10^{-7}$	1,2
	2.5×10^{-8}	5

TABLE 2. Temperature dependence of some physical properties of N-methylacetamide.

Temperature, °C	Density, g cm^{-3}	Viscosity, cP	Relative Permittivity
30.0	0.9503 (Ref. 1)	3.885 (Ref. 1)	178.9 (Ref. 1)
			179.0 (Ref. 8)
30.5			175.7 (Ref. 6)
32.5			172.9 (Ref. 6)
35.0	0.9460 (Ref. 7)	3.23 (Ref. 7)	169.7 (Ref. 6)
37.5			166.1 (Ref. 6)
40.0	0.9421 (Ref. 1)	3.019 (Ref. 1)	162.7 (Ref. 6)
			165.5 (Ref. 1)
			165.0 (Ref. 8)
			184.3[a] (Ref. 4)
45.0	0.9376 (Ref. 7)	2.54 (Ref. 7)	156.0 (Ref. 7)
50.0	0.9336 (Ref. 1)	2.411 (Ref. 1)	151.8 (Ref. 1)
60.0	0.9254 (Ref. 1)	1.958 (Ref. 1)	138.6 (Ref. 1)
			139.5 (Ref. 8)
80.0			118.5 (Ref. 8)
100			101.0 (Ref. 8)
120			87.2 (Ref. 8)
140			75.5 (Ref. 8)
160			65.0 (Ref. 8)
180			56.0 (Ref. 8)
200			48.0 (Ref. 8)

[a] Decreased rapidly with time.

Mizushima et al. (Ref. 9) have determined the structure of NMA by studying its Raman, infrared and ultraviolet spectra and by making dipole moment measurements on dilute solutions of NMA in carbon tetrachloride and in dioxane. All experimental evidence and theoretical calculations lead to the conclusion that the molecules of NMA in the liquid state exist in the trans rather than the cis configuration and that there is extensive intermolecular hydrogen bonding through which "polymers" of chain structure are formed. These structures are shown in the following figure.

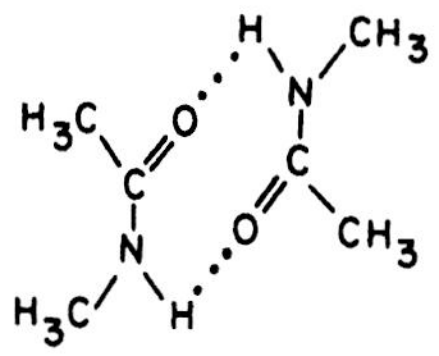

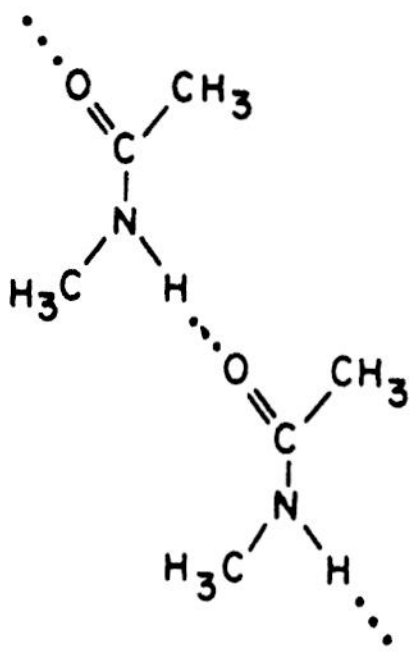

cis form (dimer) trans form ("liquid polymer")

The unusually high relative permittivity of NMA is explained on the basis of this "liquid polymer" structure. Studies of various types by other workers (Ref. 6,10-12) substantiate this conclusion regarding the structure of NMA in the liquid state.

The crystal structure and polymorphism of solid NMA has been studied by Katz (Ref. 13).

NMA is completely miscible with water; however, the value of water-NMA mixtures as solvent media is doubtful owing to the hydrolysis of the amide, particularly in the presence of bases and, to a lesser extent, acids.

Pure NMA is not very hygroscopic; however, it does take up moisture slowly if exposed to a humid atmosphere. Since NMA is a nearly neutral solvent, it does not absorb carbon dioxide from the air rapidly; however, the solubility of carbon dioxide in this solvent has not been determined. Oxygen is appreciably soluble in NMA and can be detected polarographically. It is easily removed, as presumably carbon dioxide is also, by purging the solvent with nitrogen. NMA apparently is stable in the presence of air for periods exceeding a month (however, see Ref. 4), but for long-term storage it is probably best to keep pure anhydrous NMA under an atmosphere of dry nitrogen and in the solid state, i.e., at room temperature or below.

NMA is not known to have any toxic properties. When pure it is colorless and virtually odorless. Because of these innocuous properties, and except for its melting temperature being inconveniently just above room temperature and its viscosity somewhat high, it is a convenient and pleasant solvent to work with.

The autoprotolysis constant of NMA has not been determined; however, some data are available which give good indications as to the basic strength of this solvent. Unfortunately the same is not true regarding its acidic strength as explained below. Polarographic, conductometric and glass electrode pH titration results of Knecht (Ref. 2 and 14) show that perchloric and hydrochloric acids are completely dissociated in NMA, that the second proton of sulfuric acid is less dissociated in NMA than in water (two breaks appear in the pH titration curve), and that acetic and salicylic acids are both weaker acids in NMA than they are in water. Conductance measurements of Knecht (Ref. 14) yield a value $K_a = 1.4 \times 10^{-4}$ (in water $K_a = 1.06 \times 10^{-3}$) for salicylic acid in NMA. Conductance measurements of Dawson (Ref. 15) yield a value $K_a = 7.3 \times 10^{-8}$ (in water $K_a = 1.75 \times 10^{-5}$) for acetic acid in NMA. These data indicate that NMA is a weaker base than water. The available data relating to the acidic strength of NMA are inconclusive, largely because strong bases cause decomposition of the solvent. Conductance measurements on weaker bases are inaccurate due to the relatively large contribution to the overall conductance by the solvent itself.

TESTS FOR PURITY

The major impurities commonly found in NMA are acetic acid, methylamine, a third impurity of unknown composition which is formed when NMA is heated strongly (e.g., during distillation at atmospheric pressure), and of course water. The following tests can be employed to detect these common impurities, the same tests undoubtedly serving as well to reveal the presence of other impurities. These tests are the following.

1. Water content by Karl Fischer titration.

2. Polarographic range. When tetraethylammonium perchlorate is used as the supporting electrolyte, a deaerated solution prepared from pure NMA has a polarographic range extending from +0.35 to -2.75 V (*versus* aqueous saturated calomel electrode) over which the residual current at the dropping mercury electrode does not exceed 0.4 microamp. Acetic acid impurity will show up as a cathodic discharge (or wave if only small amounts are present) in the region -1.7 to -1.9 V. Methylamine impurity will result in an anodic discharge (or wave) at potentials less positive than +0.35 V. The impurity of unknown composition mentioned above gives rise to a cathodic discharge beginning at about -2.2 V. Water as an impurity is undetected polarographically.

3. Specific conductance. It is relatively easy to prepare NMA that has a specific conductance of 1 to 5 x 10^{-7} ohm^{-1} cm^{-1}. The lowest value reported is 2.5 x 10^{-8} ohm^{-1} cm^{-1} (Ref. 5).

4. Melting temperature. The melting temperature of the solvent may be taken as an indication of purity; however, this test is less sensitive than those described above. The highest melting temperature reportedly achieved is 30.6° (Ref. 3).

5. Color and odor. Pure NMA is both colorless and odorless (or nearly odorless). Each of the common impurities (acetic acid, methylamine and the impurity of unknown composition) imparts a distinctive odor to the solvent, even when present in relatively small amounts.

PREPARATION OF NMA

NMA is available commercially (e.g., in the USA from Eastman Kodak); however, the commercial product not only is expensive but also invariably contains sufficiently large amounts of impurities (principally acetic acid) to make it unsuitable for any careful studies without purification. Since NMA is easily prepared from inexpensive starting materials, the recommended procedure is to prepare and purify the solvent as described below.

Several methods have been used by various workers (Ref. 16 and 17) to prepare NMA, but none is as simple, inexpensive and satisfactory as the direct reaction between glacial acetic acid and monomethylamine. The methylamine may be in gaseous form, in which case it is simply bubbled through hot glacial acetic acid, or in a concentrated aqueous solution. The resulting partially aqueous solution of methylammonium acetate is heated to approximately 130°C to crack out and distill off water, leaving behind a crude product consisting of NMA contaminated with relatively large amounts of acetic acid, methylamine and water.

PURIFICATION OF NMA

NMA is difficult to purify. Methods for its purification are of two general types: (a) fractional freezing, including zone melting, and (b) chemical treatment followed by distillation at reduced pressure. Both methods require more than a single cycle. That is, in the first method several (at least five) fractional freezing cycles must be employed, while in the second method several vacuum distillations, separated by different chemical treatment steps, are required. Both methods, properly handled, yield products whose purities are indistinguishable by the tests described above. [Exceptions are the solvent for which a lower specific conductance was obtained by carrying it through several additional (i.e., more than five) fractional freezing cycles, and the solvent obtained by zone melting.] However, the fractional freezing and zone melting procedures are strongly recommended for the following reasons: (a) they can readily be adapted to the purification of large volumes of solvent, (b) the equipment, at least in the case of simple fractional freezing, is minimal, easily obtained and uncomplicated, (c) they require a minimum of operator attention, (d) the process is simple and therefore not susceptible to the experimental chemical difficulties encountered in the treatment-distillation procedure, and (e) the number of cycles can be extended conveniently to obtain solvent of very low specific conductance.

Fractional freezing and zone refining procedures (recommended). Dawson and his co-workers have carried out numerous studies (Ref. 1,5,11,15,18-27) using NMA as a solvent and have developed and refined the fractional freezing technique for purifying the solvent. Their refined procedure is outlined below.

The crude solvent, from which water and much excess acetic acid have been distilled, is fractionally distilled at atmospheric pressure. Five liters of the product of this distillation (fraction collected at 204° to 206°C ?) are introduced into a six-liter Pyrex separatory funnel equipped with an inner glass finger tube which is concentric with the axis of the funnel. During the freezing process, an internal assembly is placed inside this finger. The heater assembly described by Dawson consists of 2.44 m of No. 24 Nichrome wire (resistance 1.66 ohm/ft) wrapped as 75 turns around 50 cm of 6-mm tubing and encased in a glass tube.

Crystallization is allowed to proceed at room temperature while 3 to 7 V is applied to the heater by means of a variable transformer, the voltage necessary depending upon the temperature of the room and the purity of the NMA. After a period of from five to nine days the remaining liquid is drawn off by suction. The solid is re-melted and the freezing repeated. Throughout the fractional freezing cycles the NMA is kept under an atmosphere of purified nitrogen.

In earlier papers (Ref. 1 and 11) Dawson and co-workers reported specific conductances of the solvent from 2×10^{-6} to $1\text{-}3 \times 10^{-7}$ ohm^{-1} cm^{-1} at 40°C. However, by extending the number of fractional freezing cycles and using the refined apparatus and procedure described above they were able to obtain solvent having a specific conductance less than 5×10^{-8} ohm^{-1} cm^{-1}, the lowest value reported being 2.5×10^{-8} ohm^{-1} cm^{-1}.

More recently Bonner and coworkers (Ref. 4) described the application of the zone melting technique to the purification of commercial NMA using apparatus specially constructed to be suited to substances melting near room temperature. Their apparatus consisted of Vycor tubes having a radius of 0.7 cm and a length of 78 cm, and ten such tubes containing more than one liter of material could be purified at one time. In order to maintain a small molten zone (usually about 1 cm), they placed cooling coils (cold water circulated) immediately above the heating wires. The zone was heated with a single loop of Nichrome wire, heated electrically, so as to just melt the sample in the tube. They employed a minimum of six complete melting cycles. Taking great care to avoid atmospheric contamination, they obtained a product having a melting temperature of 30.35 to 30.55°C and a relative permittivity of 184.3 at 40°C. This remarkably high relative permittivity decreased with time, however, and this behavior, coupled with results of precise calorimetric measurements and melting curves, are interpreted as an indication of some decomposition of the material, even in the solid state. They concluded that appreciable changes occur in the hydrogen bonding of the solvent when small quantities of impurities are present. The degree of purity achieved by these authors most probably is higher than any previously reported (direct comparisons cannot be made since they did not employ any of the tests for purity described above); however, the unavoidable (inherent) instability of the initial product renders this degree of purity an unattainable goal for most practical purposes. Nevertheless, zone melting, being a further refinement of the fractional freezing technique, should probably be given precedence over the simple fractional freezing technique outlined above. Assuming the availability of suitable equipment, the zone refining method can be made less time-consuming and less demanding of operator attention. The purified material can be handled with greater ease and less danger of contamination.

Other purification procedures. Fractional distillation alone at atmospheric pressure does not yield pure NMA, the product being contaminated with acetic acid, methylamine and an impurity of unknown composition which produces a cathodic polarographic discharge at about -2.2 V (*versus* aq. SCE). Despite the large difference in boiling temperature between NMA on the one hand and acetic acid and methylamine on the other, these two impurities persist even in the fraction distilling at 202 to 204°C. The impurity of unknown composition forms whenever NMA is heated to its normal boiling temperature.

Knecht and Kolthoff (Ref. 2) developed two different purification procedures involving chemical treatment steps between successive vacuum distillations. The chemical treatment steps, designed to remove the impurities mentioned above, include extraction of aqueous NMA with petroleum ether, treatment with sulfuric acid, treatment with potassium carbonate, and treatment with calcium oxide. Although both purification procedures yield pure NMA, neither procedure is recommended because both are more difficult, troublesome and time-consuming than the fractional freezing or zone melting procedures. The description of these two procedures nevertheless provides insight into the chemical properties and stability of NMA (see Ref. 2).

Attempts to remove water from NMA (even otherwise pure NMA) by treatment with phosphorus pentoxide followed by vacuum distillation yielded an impure product, the impurity giving rise to a cathodic polarographic discharge or wave beginning at -2.2 to -2.4 V. Nevertheless, Reynolds and Weiss (Ref. 28) used solvent prepared in this way for studies requiring extremely anhydrous conditions; the added impurity presumably did not interfere with their spectral measurements. French and Glover (Ref. 7) also purified NMA in this way for their conductance studies. Their procedure was to shake the solvent (presumably a commercial product) with phosphorus pentoxide, filter it through glass wool and vacuum distill it. This was repeated three times followed by two more vacuum distillations without previous treatment with phosphorus pentoxide. Their final product had a specific conductance of 4.2 to 6.0×10^{-7} ohm^{-1} cm^{-1} at 35°C which remained constant with time. Apparently they did not test polarographically for impurities, and it is probable that the impurity mentioned above was present in their solvent.

Distillation of NMA at atmospheric pressure in the presence of calcium oxide results in rapid decomposition of the solvent with the liberation of large amounts of ammonia or methylamine.

No ion-exchange or chromatographic methods have been employed for the purification of NMA; however, to purify relatively large amounts of a viscous solvent like NMA by these techniques, even if successful, probably would prove to be more difficult and time-consuming than the procedures mentioned above.

REFERENCES

1. L. R. Dawson, P. G. Sears and R. H. Graves, J. Am. Chem. Soc. 77, 1986 (1955).
2. L. A. Knecht and I. M. Kolthoff, Inorg. Chem. 1, 195 (1962).
3. L. R. Dawson, private communication.
4. O. D. Bonner, C. F. Jordan and K. W. Bunzl, J. Phys. Chem. 68, 2450 (1964).
5. L. R. Dawson, R. C. Sheridan and H. C. Eckstrom, J. Phys. Chem. 65, 1829 (1961).
6. G. R. Leader and J. F. Gormley, J. Am. Chem. Soc. 73, 5731 (1951).
7. C. M. French and K. H. Glover, Trans. Faraday Soc. 51, 1427 (1955).
8. R. Lin and W. Bannhauser, J. Phys. Chem. 67, 1805 (1963).
9. S. Mizushima, T. Simanouti, S. Nakakura, K. Kuratani, M. Tsuboi, H. Baba and O. Fujioka, J. Am. Chem. Soc. 72, 3490 (1950).
10. W. D. Kumler and C. W. Porter, J. Am. Chem. Soc. 56, 2549 (1934).
11. W. D. Williams, J. A. Ellard and L. R. Dawson, J. Am. Chem. Soc. 79, 4652 (1957).
12. J. W. Vaughn and P. G. Sears, J. Phys. Chem. 62, 183 (1958).
13. J. L. Katz and B. Post, Acta Cryst. 13, 624 (1960).
14. L. A. Knecht, unpublished data.
15. L. R. Dawson, J. W. Vaughn, M. E. Pruitt and H. C. Eckstrom, J. Phys. Chem. 66, 2684 (1962).
16. G. F. D'Alelio and E. E. Reid, J. Am. Chem. Soc. 59, 109 (1937).
17. A. Berger, A. Loewenstein and S. Meiboom, J. Am. Chem. Soc. 81, 62 (1959).
18. L. R. Dawson, E. D. Wilhoit and P. G. Sears, J. Am. Chem. Soc. 78, 1569 (1956).
19. L. R. Dawson, E. D. Wilhoit, R. R. Holmes and P. G. Sears, J. Am. Chem. Soc. 79, 3004 (1957).
20. L. R. Dawson, G. R. Lester and P. G. Sears, J. Am. Chem. Soc. 80, 4233 (1958).
21. L. R. Dawson and W. W. Wharton, J. Electrochem. Soc. 107, 710 (1960).
22. L. R. Dawson, J. E. Berger and H. C. Eckstrom, J. Phys. Chem. 65, 986 (1961).
23. L. R. Dawson, J. W. Vaughn, G. R. Lester, M. E. Pruitt and P. G. Sears, J. Phys. Chem. 67, 278 (1963).
24. L. R. Dawson, J. E. Berger, J. W. Vaughn and H. C. Eckstrom, J. Phys. Chem. 67, 281 (1963).
25. H. C. Eckstrom, D. G. Oei and L. R. Dawson, J. Phys. Chem. 67, 2804 (1963).
26. L. R. Dawson, W. H. Zuber and H. C. Eckstrom, J. Phys. Chem. 69, 1335 (1965).
27. L. R. Dawson and D. G. Oei, J. Colloid Sci. 20, 282 (1965).
28. W. L. Reynolds and R. H. Weiss, J. Am. Chem. Soc. 81, 1790 (1959).

N-Methylpropionamide

Prepared for publication by

T. B. Hoover

N-Methylpropionamide (abbreviated NMP in this report) is a solvent remarkable chiefly for its high relative permittivity (dielectric constant). At -40°C, the relative permittivity of the undercooled liquid is 341 (Ref. 1), one of the highest values known for a nonionic liquid. This property, together with a wide liquid range (normal boiling temperature 193°C: Ref. 2) and good chemical and thermal stablity, make NMP an interesting and valuable solvent for electrochemical research. Its chemical properties resemble those of N-methylacetamide (Ref. 3 and Chapter 11).

PHYSICAL PROPERTIES OF NMP

Pure NMP is a colorless, odorless liquid at room temperature. Apparently no toxic properties have been reported although, in common with several other N-alkyl aliphatic amides, NMP was teratogenic in rats (Ref. 4). Teratogenicity was enhanced by sodium nitrite supplied in drinking water with the NMP (Ref. 5). In Table 1 some physical properties are listed according to the source of the data, since there were evident differences in the purity of the NMP samples on which the properties were measured. Some reported values for the boiling temperature are 60°C/ 0.5 torr (Ref. 6), 103°C/12-13 torr (Ref. 7), 134°C/60 torr (Ref. 8), and 146°C/90 torr (Ref. 9) (1 torr = 133.32 N m^{-2}). The vapor pressures measured by Gopal and Rizvi (Ref. 2) lead to an extrapolated normal boiling temperature of 193°C. The vapor pressure data correspond to somewhat lower boiling temperatures at reduced pressures than those given by the preceding references.

Estimates of the high-frequency relative permittivity ε_∞, derived according to Lindenberg's formula (Ref. 10) from the gas-phase dipole moment of NMP and its static relative permittivity, are appreciably higher than estimates derived from the refractive index (Ref. 1). Static relative permittivities were measured over the range -40°C to 120°C by Cole and co-workers (Ref. 1) with a transformer bridge at 100 kHz. The sample they used appears to have been impure, however, on the basis of the low freezing temperature and reported corrosion of the nickel cell. Malmberg and Hoover (Ref. 11 and 12) also measured static relative permittivities from the freezing temperature (-30°C) to 150°C. These data were obtained at audio-frequencies both with a transformer bridge and with a Wheatstone conductivity bridge. Several preparations of NMP were measured in guarded and in two-terminal cells. The combined results of Malmberg and Hoover (Ref. 12) are represented by either of the following equations with a standard deviation less than 0.5 unit in ε.

$$\varepsilon = 231.6 - 2.7207\, t + 0.02123\, t^2 - 1.098 \times 10^{-4}\, t^3 + 2.534 \times 10^{-7}\, t^4$$

$$\log \varepsilon = 2.3649 - 5.0427 \times 10^{-3}\, t + 9.8455 \times 10^{-6}\, t^2 - 1.915 \times 10^{-8}\, t^3$$

The estimated error of the measured values is within two per cent. Densities of the same specimens of NMP were measured over the range -25°C to 150°C in four pyknometers and quartz dilatometers (Ref. 12). With a standard deviation of 5 x 10^{-5}, the densities (g cm^{-3}) were represented by the equation

$$\rho = 0.95069 - 8.136 \times 10^{-4}\, t + 2.921 \times 10^{-7}\, t^2 - 1.79 \times 10^{-9}\, t^3$$

The dipole moment has been reported (Ref. 13) to be 3.59 D.

Della Monica (Ref. 14) reported viscosities over a wide temperature range although the corresponding values for density and specific conductance were not typical of a pure sample.

From the vapor pressure measurements (Ref. 2) the calculated heat of vaporization is 54.4 kJ mol^{-1}. The cryoscopic constant is 4.7 K kg mol^{-1} on the basis of the writer's measurements with cyclohexane and N-methylacetamide as solutes (Ref. 15). The corresponding heat of fusion is 9.2 kJ mol^{-1}.

TABLE 1. Physical properties of N-methylpropionamide

Property	Ref. 8	Ref. 11	Ref. 12	Ref. 14	Ref. 26	Ref. 27	Ref. 29	Ref. 35
Freezing temperature, °C		-30.9				-30.9		
Specific conductance at 25°C, $ohm^{-1}\ cm^{-1} \times 10^6$	1	0.12		7.17		0.22	0.6	
Refractive index, n_D^{25}	1.4350	1.4350						
Density, $g\ cm^{-3}$								
25°C	0.9306		0.93050	0.94465		0.9308	0.9305	
30°C		0.92647	0.92650		0.9269			
40°C		0.91847	0.91850		0.9188			
Viscosity, cP*								
25°C				5.2005		5.25	5.21	5.215
30°C					4.568	4.58		4.554
35°C				3.9920		3.56		3.533
40°C					3.541			
45°C				3.1055				
60°C				2.2378				
80°C				1.1552				
Relative permittivity, ε								
20°C	179.8	185.3						
25°C	172.2		175.4			176	176	
30°C	164.3	166.7			164.3			
40°C		150.6	150.7		148.9			
60°C			124.2		117.9			

*1 cP = $1.0 \times 10^{-3}\ N\ s\ m^{-2}$

CHEMICAL PROPERTIES OF NMP

NMP is essentially neutral and inert as a solvent but there is almost no quantitative information on its acidity or basicity. Nuclear magnetic resonance studies (Ref. 16) showed that NMP was 50 per cent protonated in an aqueous sulfuric acid solution having an acidity value $H_A = -0.95$. H_A is a special acidity function of Yates et al. (Ref. 17) that was shown to be applicable to amides. At the value in question it was substantially the same as the Hammett H_0. This H_A scale has been revised by Edward and Wong, however (Ref. 18). Thermochemical measurements (Ref. 19) showed a greater exothermic heat of solution for hydrogen chloride in NMP than in water, partly as a result of the difference in relative permittivities. On the basis of precise conductance measurements (Ref. 6), hydrogen chloride is a strong electrolyte in NMP, having a limiting molar conductance and ionization comparable to those of the alkali halides. The specific conductance of purified NMP suggests that the autoprotolysis constant probably is less than 5×10^{-11} M^2, estimated as 10^6 κ^2/Λ_0^2. Rapid equilibration of NMP with D_2O indicated that the amide H is labile (Ref. 20) but not necessarily ionizable. Protolysis of NMP is catalysed by acid and base, and the kinetics have been measured at 34°C in water (Ref. 21) and in 16 mol-per-cent tert.-butyl alcohol (Ref. 22). The respective values of the base-catalysed rate constant, k_{OH}, are 3.8×10^6 and 2.8×10^5 mol^{-1} s^{-1} and of the acid-catalysed constant, k_H, 1.1×10^3 and 6.3×10^2 mol^{-1} s^{-1}. The activation energies for the protolysis in aqueous solution were 46.3 kJ mol^{-1} for the base-catalysed reaction and 35.8 kJ mol^{-1} for the acid-catalysed protolysis. Electrolysis of NMP containing 5-10 per cent sulfuric acid yielded the methane bis-amide (4,6-diaza-3,7-nonadione), among other products (Ref. 23).

The kinetics of the formation and hydrolysis of NMP in dilute aqueous solution have been measured (Ref. 24). Although NMP is thermodynamically unstable with respect to its hydrolysis products, the rate of hydrolysis is slow. The second-order rate constant for the OH^--catalysed hydrolysis in dilute solution, extrapolated to 25°C, is 3.13×10^{-6} l mol^{-1} s^{-1} (Ref. 24). Aqueous NMP has been used successfully as a high relative permittivity solvent for studying the kinetics of hydrolytic reactions (Ref. 25). In a study of the hydrolysis of alkyl halides, Ross and Labes (Ref. 7) could not determine whether NMP was electrophilic, nucleophilic, or both.

As is to be expected from its high relative permittivity, NMP is a good electrolytic solvent. Conductivity measurements have shown that most electrolytes are highly ionized at much greater concentrations than in water (Ref. 6, 26-29). Nevertheless, distinct evidence of ion-pair formation has been found (Ref. 6 and 27). Apparently NMP has not been used as a polarographic solvent, but the double-layer capacity at the mercury drop was measured for 0.1 M KPF_6 in NMP (Ref. 30). The point of zero charge was -0.342 V versus the normal calomel electrode. Two capacity maxima (Ref. 30), on either side of the zero charge potential, probably represent different orientations of the solvent dipoles in the double layer. The platinum-hydrogen electrode and the silver-silver chloride electrodes behave reversibly in NMP (Ref. 31 and 32).

PREPARATION AND PURIFICATION OF NMP

NMP is available commercially (e.g., in the USA, from Eastman Organic Chemicals). One fractional vacuum distillation of Eastman NMP yielded a product of high quality (Ref. 6 and 33). Synthesis from inexpensive starting materials is easy but purification of the product is quite laborious. NMP has been made by the reaction of methylamine with propionyl chloride (Ref. 9, 29 and 34) but, for kilogram quantities, it is probably preferable to start with propionic acid (Ref. 8 and 26). The following procedure is recommended.

Pass anhydrous methylamine rapidly into well-stirred, 50 per cent excess propionic acid (or add an aqueous solution of methylamine to the theoretical quantity of acid: Ref. 1). Heat the mixture rapidly to 120-140°C, taking off water through a simple distilling head. Since hydrolysis occurs during this operation, the more rapidly the water can be removed the better the yield. Distillation of the xylene azeotrope has also been used to remove water (Ref. 26). Purify the crude NMP by fractional distillation at 5-15 torr. Although the expected impurities are more volatile than NMP, removal is difficult and the distillation often must be repeated several times in order to obtain a product of high quality (Ref. 27,30). Although several workers have treated the crude NMP with calcium oxide before or during the distillation (Ref. 1,2,7 and 33), there appears to be little benefit from this procedure (Ref. 30). If there are no chemical additives in the pot, it appears that no volatile impurities are present or generated that have higher boiling points than NMP. Consequently, the quality of the distillate, as judged by its specific conductance (Ref. 27), increases during the distillation. Several pot residues can be combined and re-distilled to yield good distillate NMP until the final residue becomes quite dark colored.

Although NMP does not seem to be subject to oxidation by the air, it can pick up appreciable moisture from the atmosphere and is best stored for extended periods over barium oxide or Molecular Sieves (Ref. 19). It is much easier to redistill solvent that has been used in conductivity measurements (and, presumably, other experiments) than to repeat the synthesis and purification.

TESTS FOR PURITY

Few specific tests for impurities in NMP have been developed or evaluated. The Karl Fischer titration of water can be performed directly in NMP solvent without dilution and has been used to detect water concentrations at the parts per million level (Ref. 7 and 11). Gas chromatography (Ref. 1,6,27 and 28) and NMR (Ref. 6) have been used to characterize NMP but only water and propionic acid have been reported as identified impurities.

The purity of NMP is usually judged by the values of its physical constants, specific conductance being one of the more sensitive tests. As Table 1 shows, a specific conductance at 25°C less than 1×10^{-6} ohm^{-1} cm^{-1} can be obtained routinely: the lowest value that has been reported (Ref. 28) is 0.8×10^{-7} ohm^{-1} cm^{-1}. This property is indicative only of total ionic impurities. It should also be borne in mind that limiting molar conductivities in NMP are about one-fifth as great as in water because of the greater viscosity of NMP. Freezing temperature lowering is a more general measure of impurity but is inconvenient to determine because of the low temperature, high viscosity of the liquid, and the great tendency of NMP to undercool. The relative permittivity is quite sensitive to water (Ref. 11), 0.1 per cent water reducing the relative permittivity of NMP about one per cent, but the effect of other likely impurities is not known. The viscosity of NMP is increased appreciably by the addition of small amounts of water (ref. 28) and most electrolytes (Ref. 28,29,35 and 36), but is lowered by some organic solutes (Ref. 35).

STRUCTURAL INTERPRETATION OF THE PROPERTIES OF NMP AND ITS SOLUTIONS

Infrared spectra (Ref. 34 and 37) and proton magnetic resonance studies (Ref. 38,39 and 40) of the structure of N-alkyl aliphatic amides show conclusively that the high relative permittivities result from chainwise hydrogen-bonded association of <u>trans</u> isomers that have parallel orientation of the individual dipole moments. Only the formamides contain a significant proportion of the <u>cis</u> form (Ref. 38), which can associate in dimers with anti-parallel moments. For substituents bulkier than methyl on both the carbonyl carbon and the nitrogen, steric requirements favor the <u>trans</u> form but reduce the extent of association. NMP, which exists almost exclusively in the <u>trans</u> form (Ref. 34,37,38 and 39), has a lower relative permittivity at 30°C than N-methylacetamide (Ref. 41) but a higher value of the Kirkwood correlation factor (Ref. 1), $\underline{g}$, which is a measure of the number of unit dipoles that act cooperatively. Proton magnetic resonance measurements (Ref. 40) confirmed the values of $\underline{g}$ for NMP derived from the relative permittivity. The effects of solutes upon this hydrogen-bonded structure are reflected in changes in the physical properties of the solutions.

At all concentrations water is a net structure-breaking agent in NMP, i.e., more NMP-NMP bonds are broken than NMP-H_2O bonds formed. This conclusion was inferred from the densities of the solutions (Ref. 11 and 42), and confirmed by nuclear magnetic resonance studies (Ref. 39). Nevertheless, the increased viscosity produced by traces of water in NMP suggests that there must be some crosslinking of the NMP chains by water although the mean chain length is shortened. Infrared spectra of the NMP-water system indicated stronger hydrogen-bonding between unlike molecules than either pair of like molecules (Ref. 43).

Organic solutes also have pronounced effects on the liquid structure of NMP. On the basis of density and viscosity measurements (Ref. 35), pyridine breaks up the amide structure, presumably through hydrogen bonding. Benzene solutions, also studied calorimetrically (Ref. 44), had less pronounced structure-breaking effects, while carbon tetrachloride, tetrachloroethylene, cyclohexane, and n-hexane showed little or no specific interaction with NMP (Ref. 44).

From the temperature coefficients of Walden products, Singh (Ref. 45,46,47 and 48) has classified Li^+, Na^+, and K^+ as net structure-reinforcing agents and Cl^-, Br^-, I^-, Rb^+, Cs^+, and most tetraalkylammonium ions as structure-breaking. The variation of viscosity B-coefficients with temperature (Ref. 36) and apparent molal volumes (Ref. 49) generally confirmed this classification. The viscosity effects of a series of tetraalkyammonium ions agreed closely with the predictions of the Einstein equation, based on their crystallographic radii (Ref. 50). This result indicates that these ions have a purely geometrical effect on NMP, in contrast to the situation in aqueous solutions. It is interesting that the limiting ionic mobility of K^+ in NMP is greater than that of Na^+ and the same as that of tetrapropylammonium ion (Ref. 33) and that the limiting mobilities of Cl^-, Br^-, and I^- increase in that order.

From conductance and viscosity data for hydrogen chloride in NMP (Ref. 51), the Walden product and ionic mobility of the proton are similar to those of K^+ (Ref. 48), and the proton viscosity B-coefficient decreases with increasing temperature comparable to those of K^+ and Li^+ (Ref. 36). These results indicate that the proton is structure-making but not markedly different from other small cations. Potentiometric studies of the dissociation of protonated tris(hydroxymethyl)aminomethane suggest that mixtures of NMP with methanol or with water are more basic than the pure liquids (Ref. 52 and 53), but do not provide an unequivocal measure of the interaction of the proton with NMP. Infrared spectroscopy (Ref. 54) indicated that Li^+ is complexed to the carbonyl oxygen of NMP. Although the variation of Walden product with temperature has been interpreted as evidence for ionic solvation (Ref. 45, 46,47 and 48), it is doubtful whether ions larger than H^+ and Li^+ are specifically solvated. The structural effects of ions upon NMP seem to be mainly electrostatic in nature (Ref. 55), although other geometric factors have been discussed with respect to tetraalkylammonium ions (Ref. 56).

REFERENCES

1. S. J. Bass, W. I Nathan, R. M. Meighan and R. H. Cole, J. Phys. Chem. 68, 509 (1964).
2. R. Gopal and S. A. Rizvi, J. Indian Chem. Soc. 45, 13 (1968).
3. L. A. Knecht, Pure Appl. Chem. 27, 283 (1971).
4. T. von Kreybig, R. Preussman and W. Schmidt, Arzneim.-Forsch. 18, 645 (1968); Chem. Abstr. 69, 42521a (1968).
5. P. Hafen V. Koblitz, R. Atzeroth, Th. Von Kreybig, Z. Krebsforsch. Klin. Onkol. 80, 269 (1973).
6. W. C. Duer, R. A. Robinson, and R. G. Bates, J. Solution Chem. 5, 765 (1976).
7. S. D. Ross and M. M. Labes, J. Amer. Chem. Soc. 79, 4155 (1957).
8. G. R. Leader and J. F. Gormley, J. Amer. Chem. Soc. 73, 5731 (1951).
9. G. F. D'Alelio and E. E. Reid, J. Amer. Chem. Soc. 59, 109 (1937).
10. A. B. Lindenberg, C. R. Acad. Sci., Paris. Ser C. 262, 1504 (1966).
11. T. B. Hoover, J. Phys. Chem. 73, 57 (1969).
12. Electrochemical Analysis Section: Summary of Activities, July 1966 to June 1967 (R. G. Bates, Ed.) NBS Tech. Note No. 423 (October 1967).
13. R. M. Meighan and R. H. Cole, J. Phys. Chem. 68, 503 (1964).
14. M. Della Monica, Electrochim. Acta. 21, 641 (1976).
15. Electrochemical Analysis Section: Summary of Activities, July 1967 to June 1968 (R. G. Bates, Ed.) NBS Tech. Note No. 453 (July 1968).
16. M. Liler, J. Chem. Soc. 1969B, 385.
17. K. Yates, J. B. Stevens and A. R. Katritsky, Canad. J. Chem. 42, 1957 (1964).
18. J. T. Edward and S. C. Wong, Can. J. Chem. 55, 2492 (1977).
19. J. L. Curnutt, Dissert. Abstr. (B), 29, 4608 (1969).
20. K. V. Ramiah, V. V. Chalapathi and C. A. I. Chary, Current Sci. (India), 35, 350 (1966).
21. L. C. Martinelli, C. D. Blanton and J. F. Whidby, J. Phys. Chem. 75, 1895 (1971).
22. L. C. Martinelli, C. D. Blanton and J. F. Whidby, J. Amer. Chem. Soc. 93, 5111 (1971).
23. D. E. Couch, Electrochim. Acta, 9, 327 (1964).
24. H. Morawetz and P. S. Otaki, J. Amer. Chem. Soc. 85, 463 (1963).
25. R. K. Wolford and R. G. Bates, J. Phys. Chem. 66, 1496 (1962).
26. L. R. Dawson, R. H. Graves and P. G. Sears, J. Amer. Chem. Soc. 79, 298 (1957).
27. T. B. Hoover, J. Phys. Chem. 68, 876 (1964).
28. T. B. Hoover, J. Phys. Chem. 68, 3003 (1964).
29. J. Sandeaux, B. Brun and J. Molenat, J. Chim. Phys. 68, 480 (1971).
30. R. Payne, J. Phys. Chem. 73, 3598 (1969).
31. E. S. Etz, R. A. Robinson and R. G. Bates, J. Solution Chem. 1, 507 (1972).
32. E. S. Etz, R. A. Robinson and R. G. Bates, J. Solution Chem. 2, 405 (1973).
33. R. Gopal and O. N. Bhatnager, J. Phys. Chem. 70, 4070 (1966).
34. R. L. Jones, Spectrochim. Acta 22, 1555 (1966).
35. F. J. Millero, J. Phys. Chem. 72, 3209 (1968).
36. R. Gopal and P. P. Rastogi, Z. Phys. Chem. (Frankfurt am Main) 69, 1 (1970).
37. R. L. Jones, Spectrochim. Acta 23A, 1745 (1967).
38. L. L. LaPlanche and M. T. Rogers, J. Amer. Chem. Soc. 86, 337 (1964).
39. J. F. Hinton and C. E. Westerman, Spectrochim. Acta 26A, 1387 (1970).
40. S. Takagi, T. Imura, and K. Kawabe, Technol. Rep. Osaka Univ. 23, 395 (1973); Chem. Abs. 81, 62953u (1974).
41. J. W. Vaughn and P. G. Sears, J. Phys. Chem. 62, 183 (1958).
42. P. Rohdewald and M. Moeldner, J. Phys. Chem. 77, 373 (1973).
43. O. D. Bonner and Y. S. Choi, Spectrochim. Acta 31A, 1975 (1975).
44. M. Lindheimer, G. Etienne, and B. Brun, J. Chim. Phys. 71, 135 (1974).
45. R. D. Singh and R. Gopal, Bull. Chem. Soc. Japan 45, 2088 (1972).
46. R. D. Singh, Bull. Chem. Soc. Japan 46, 14 (1973).
47. R. D. Singh, Z. Phys. Chem. (Frankfurt am Main) 90, 34 (1974).
48. R. D. Singh and M. M. Husain, Z. Phys. Chem. (Frankfurt am Main) 94, 193 (1975).
49. R. Gopal and K. Singh, Z. Phys. Chem. (Frankfurt am Main) 91, 98 (1974).
50. P. P. Rastogi, Z. Phys. Chem. (Frankfurt am Main) 86, 43 (1973).
51. J. E. Van Evercooren, G. V. Merken, and H. P. Thun, Bull. Soc. Chim. Belg. 84, 533 (1975).
52. R. G. Bates, J. S. Falcone, Jr., and A. Y. W. Ho, Anal. Chem. 46, 2004 (1974).
53. Y. Bokra and R. G. Bates, Anal. Chem. 47, 1110 (1975).
54. A. F. Diorio, E. Lippincott, and L. Mandelkern, Nature, London 195, 1296 (1962).
55. P. P. Rastogi, Z. Phys. Chem. (Frankfurt am Main) 75, 202 (1971).
56. K. Singh, D. K. Agarwal, and R. Kumar, J. Indian Chem. Soc. 52, 304 (1975).

Printed in France by Amazon
Brétigny-sur-Orge, FR

51786473R00040